STANLEY TIFFANY

THE MEAD MAKING BIBLE

" You Catch More Flies With Honey Than Vinegar "

Copyright © 2024 by Stanley Tiffany

All rights reserved. No part of this publication may be reproduced, stored or transmitted in any form or by any means, electronic, mechanical, photocopying, recording, scanning, or otherwise without written permission from the publisher. It is illegal to copy this book, post it to a website, or distribute it by any other means without permission.

First edition

This book was professionally typeset on Reedsy.
Find out more at reedsy.com

Contents

INTRODUCTION	vii
I INTRODUCTION TO MEAD	
1 CHAPTER ONE	3
THE HISTORY AND CULTURE OF MEAD	3
GLOBAL MEAD TRADITIONS	9
The Revival Of Mead In Modern Times	12
2 CHAPTER TWO	15
UNDERSTANDING MEAD	15
What Is Mead?	15
Ingredients Breakdown For Mead Making	16
Honey	16
Water	22
Yeast	26
Additional Fermentables (Fruits,Spices,Syrups)	29
Clarifying Agents	33
3 CHAPTER THREE	37
TYPES OF MEAD	37
II THE MEAD MAKING PROCESS	
4 CHAPTER FOUR	49
GETTING STARTED	49
Essential Equipment and Tools	49
Sanitation: Importance and Methods	54

Setting Up Your Mead Making Space	59
5 CHAPTER FIVE	60
PREPARING YOUR MUST	60
Measuring and Mixing Ingredients	64
Understanding the Role of pH and Acidity	67
Adjusting Sugar Levels: Hydrometer Use and Potential Alcohol Content	71
Adding Nutrients: Importance of Yeast Nutrients and Feeding Schedules	76
6 CHAPTER SIX	81
FERMENTATION	81
Primary vs. Secondary Fermentation: Processes and Differences	82
Managing Fermentation: Temperature, Duration, and Signs of Progress	86
Troubleshooting Common Fermentation Issues	89
RACKING	93
Techniques and Equipment for Racking	95
Impact of Racking on Mead Quality	97
When to Transfer Mead for Aging	98
How to Transfer Mead for Aging	99
7 CHAPTER SEVEN	101
AGING AND MATURATION	101
The Importance of Aging in Mead	102
Aging Vessels	104
Types of Aging Vessels	105
Impact of Aging Vessels on Maturation	106
Monitoring and Tasting During Aging	107
Adjusting Flavors Post-Fermentation	111
Clarifying Mead	115

III   STORAGE AND ADVANCED MEAD MAKING TECHNIQUES

8 CHAPTER EIGHT	121
BOTTLING AND STORAGE	121
Choosing Bottles and Closures	122
Proper Bottling Techniques	126
Labeling and Record-Keeping	130
9 CHAPTER NINE	133
EXPERIMENTING WITH FLAVORS	133
Incorporating Fruits: Techniques and Timing	134
Using Spices and Herbs	137
Specialty Meads: Creating Metheglins, Melomels, and Others.	140
10 CHAPTER TEN	145
SCALING UP PRODUCTION	145
Managing Fermentation in Bulk	149
11 CHAPTER ELEVEN	152
TROUBLESHOOTING AND REFINING	152
Refining Techniques: Filtration, Fining, and Stabilization	156
Experimentation and Iteration	160

IV   MEAD IN CULTURE AND COMMERCE

12 CHAPTER TWELVE	165
MEAD IN MODERN CULTURE	165
13 CHAPTER THIRTEEN	168
MEAD AND LEGAL CONSIDERATIONS	168
Understanding Local Laws: Homebrewing vs. Commercial Production	171
Licensing and Regulations for Commercial Mead Makers	177
Navigating Taxation and Distribution	181
14 CHAPTER FOURTEEN	185

	Planning and Strategies	186
	Scaling Up: Managing Growth and Expansion	190

V MEAD RECIPES

15 CHAPTER FIFTEEN — 197
  TRADITIONAL MEAD RECIPES — 197
  - Basic Traditional Mead — 201
  - Orange Blossom Mead — 204
  - Acacia Honey Mead — 207

16 CHAPTER SIXTEEN — 212
  FRUITS AND SPICE MEADS — 212
  - Blackberry Melomel — 214
  - Hibiscus and Ginger Metheglin — 216

17 CHAPTER SEVENTEEN — 220
  EXPERIMENTAL MEAD — 220
  - Coffee and Vanilla Mead — 221
  - Chocolate Cherry Bochet — 224
  - Lavender Lemon Balm Mead — 228

18 CHAPTER EIGHTEEN — 233
  SEASONAL AND HOLIDAY MEADS — 233
  - Autumn Spice Mead — 234
  - Winter Solstice Mulled Mead — 236
  - Spring Blossom Mead — 239

19 CHAPTER NINETEEN — 242
  MEAD PAIRING AND COCKTAILS — 242
  - Pairing Mead with Cheese, Meat, and Dessert — 243
  - Mead-Based Cocktails — 247
  - Crafting Mead-Based Aperitifs and Digestifs — 252

20 CONCLUSION. — 258

INTRODUCTION

Welcome to the world of mead, where ancient tradition meets modern creativity! If you're holding *THE MEAD MAKING BIBLE*, I can already tell that you're someone with an adventurous spirit. Maybe you've heard whispers of this golden drink of the gods—smooth, sweet, complex—and you're curious to know more. Or perhaps, like me, you've been captivated by the idea of transforming simple ingredients into something truly magical. Either way, you're about to embark on a journey as old as civilization itself, one that will take you deep into the heart of one of the oldest fermented drinks known to humankind.

I remember when I first started making mead. The process seemed mysterious—honey, water, yeast—how could something so simple result in such a range of flavors and styles? But as I began experimenting, I discovered that mead is a drink that evolves with you. It's forgiving to the novice and rewarding to the expert, and there's always something new to learn. Whether you're here to make a classic traditional mead or to push the boundaries with bold flavor combinations, this book will give you all the tools and insights you need to craft mead with confidence.

In *THE MEAD MAKING BIBLE*, you'll find detailed guide on everything from selecting the best honey to perfecting fermentation techniques. We'll start at the very beginning—what exactly mead is, how it differs from other alcoholic beverages, and why it has such a special place in history. You'll quickly see why mead was once considered the "nectar of the gods," a drink revered by cultures across the world, from the Vikings of Scandinavia to the monks of medieval Europe. And now, you'll have the knowledge to create your own mead at home, one that's just as divine.

Here are few topics that will be found in this book;

1) THE HISTORY AND CULTURE OF MEAD

2) TRADITIONAL MEAD RECIPES

3) FERMENTATION

4) STARTING A MEADERY AND SCALING UP YOUR MEAD PRODUCTION

5) AGING AND MATURATION

6) TROUBLESHOOTING AND REFINING

As you explore these pages, you'll discover that making mead is both an art and a science. I'll walk you through the importance of choosing the right honey—whether it's wildflower honey for a light, floral mead or buckwheat honey for a dark, rich flavor. We'll dive into the role of water, often overlooked but essential to creating a clean, balanced mead. You'll get to know yeast like an old friend, understanding which strains will produce a sweet, delicate mead, and which will deliver something bold and high in alcohol. And beyond the basics, we'll experiment with fruits, spices, herbs, and even hops to create meads that are uniquely yours.

But that's just the beginning. You'll also learn how to age your mead for optimal flavor, troubleshoot common issues, and adjust your brew after fermentation to get it just right. Whether you're aiming for a crisp, refreshing mead to enjoy after a long day or a robust, complex bottle to lay down and age, you'll find plenty of tips and tricks to make every batch a success.

Throughout the book, you'll find recipes ranging from simple, no-fuss traditional meads to more elaborate concoctions like melomels (meads with fruit), metheglins (meads with spices and herbs), and even sparkling meads for those who want a bit of effervescence. No matter your skill level, there's something for everyone, and I'll be with you every step of the way—sharing stories, offering advice, and most importantly, helping you unlock the full potential of mead-making.

So, what can you look forward to as you dive in? Expect to gain a deep understanding of the ingredients and processes that make mead-making so rewarding. You'll learn how to tweak recipes to suit your personal taste, whether you prefer a dry, crisp finish or a sweet, honey-forward profile. You'll also develop the confidence to try new things—perhaps adding seasonal fruits to your next batch or experimenting with aging techniques to bring out hidden complexities.

And here's the best part—once you've made your first few batches of mead, you'll realize how limitless the possibilities are. The world of mead is vast, with endless variations just waiting for your creative touch. By the end of this book, you'll not only be a mead-maker—you'll be a mead artist, capable of crafting unique and delicious brews that reflect your own personal style.

So, get ready. Gather your ingredients, clear some space in your kitchen, and prepare to join the ranks of mead-makers from across the centuries. Whether you're brewing for yourself, impressing friends, or even dreaming of turning this into a full-fledged hobby, *THE MEAD MAKING BIBLE* is here to guide you through every exciting, sticky, and delicious step.

Let's get started! Your mead-making adventure awaits.

I

INTRODUCTION TO MEAD

1

CHAPTER ONE

THE HISTORY AND CULTURE OF MEAD

Mead, also known as honey wine, is an alcoholic drink made by fermenting honey mixed with water. Sometimes, fruits, spices, grains, or hops are added for flavor. The alcohol content can range from around 3.5% to over 20%. Mead is one of the oldest alcoholic drinks in history, and what makes it unique is that most of the sugar used in fermentation comes from honey. Mead can be still, carbonated, or naturally sparkling, and it doesn't have to be sweet—it can also be dry or semi-sweet. Mead is often considered the world's oldest alcoholic beverage. Its origins can be traced back to 7000 B.C. in Northern China, where traces of fermented honey, rice, and fruit were found in pottery shards. This ancient brew indicates that early humans discovered the natural fermentation of honey, likely by accident when wild honey mixed with water and yeasts.

Mead made with spices is called metheglin, and mead made with fruit is called melomel. While people sometimes use the term "honey wine" as another name for mead, wine usually refers to drinks made from fermented grapes or other fruits. In some cultures, "honey wine" refers to a different drink altogether. For example, in Hungary, their honey wine is made by fermenting honey with the leftover pulp from grapes or other fruits.

Mead's ancient origins suggest that it was likely discovered by accident, long before humans had mastered agriculture or fermentation. Early humans, gathering honey from wild bees, may have inadvertently left it exposed to rain, causing the honey to mix with water. Natural yeasts in the environment would have initiated the fermentation process, transforming the sweet liquid into a mildly alcoholic beverage. This serendipitous discovery would have been nothing short of miraculous to early humans, who had limited means of preserving food and drinks.

As people began to understand and refine the process, mead became a staple in many ancient cultures, predating even the brewing of beer and the fermenting of wine. Its production didn't require advanced tools or techniques, making it accessible across various societies, from nomadic tribes to settled agricultural communities. This widespread presence is a testament to its versatility and the universal appeal of its simple ingredients.

The earliest written mention of mead might be found in the hymns of the Rigveda, a sacred text of the ancient Vedic religion and later Hinduism, dating back to around 1700–1100 BCE. Since the Rigveda predates the split between the Indo-Iranian peoples (around 2000 BCE), this reference to mead could trace back to the Western Steppe or Eastern Europe. The Abri, a northern group of the Taulantii tribe, were known by ancient Greek writers for their method of making mead from honey. In ancient Greece's Golden Age, mead was said to be the favored drink. Aristotle discussed mead made in Illiria, and Pliny the Elder mentioned it in his work, *Naturalis Historia*, where he distinguished mead from honey-sweetened wine. The Roman writer Columella even provided a recipe for mead in his work *De re rustica*, around 60 CE.

This shows that mead has deep roots in ancient cultures, valued and recorded by some of history's most influential civilizations.

For making mead during this period, They make use of rainwater that has been stored for several years and mix one part of this water with a pound of honey. For a lighter mead, they mix one part water with nine ounces of honey. Leave the mixture in the sun for 40 days, then place it near a fire. If they don't have rainwater, they boil spring water instead.

The ancient Greek writer Pytheas mentioned coming across a drink made from grain and honey, similar to mead, during his travels in Thule (an ancient name for a distant northern land, possibly in modern Scandinavia). Historian James Henry Ramsay suggested that this drink was an early version of Welsh metheglin, a spiced mead.

When Prince Charles II, then just 12 years old, visited Wales in 1642, Welsh metheglin was served at a feast. This wasn't just a tasty drink; it was a way for the Welsh to assert their cultural identity during a time when the idea of a unified British identity was still forming. This period was marked by the Union of the Crowns in 1603, when England and Scotland were ruled by the same monarch, and eventually led to the creation of the Kingdom of Great Britain in 1707.

Welsh metheglin at that feast symbolized the rich and unique traditions of Wales, reminding everyone that Welsh culture was a significant part of the emerging British identity. It also shows how mead was not just a drink but a cultural symbol used to express national pride.

A poem called *Kanu y med*, or "Song of Mead" (*Cân y medd* in Welsh), is attributed to the Welsh bard Taliesin, who lived around 550 CE. This poem celebrates the importance of mead in the culture of the time. The image of warriors drinking, feasting, and boasting in mead halls is a powerful one that appears not just in Welsh tradition but also in other early European cultures.

For example, the poem *Y Gododdin*, attributed to the poet Aneirin, who was likely a contemporary of Taliesin, describes a mead hall in Din Eidyn (now modern-day Edinburgh). In this hall, warriors gathered to drink mead before heading into battle. This scene is echoed in the Old English epic

Beowulf, where Danish warriors also drank mead. These mead halls were more than just places to drink; they were centers of social life, where stories were told, alliances were formed, and heroes were celebrated.

In both Insular Celtic and Germanic poetry, mead is depicted as the drink of choice for heroes and even gods. This wasn't just a beverage; it symbolized the camaraderie and valor of the warrior class. Mead was a key part of the cultural fabric, representing not just physical sustenance but also the strength and unity of those who shared it.

In medieval Ireland, mead, known in Old Irish as *mid*, was a beloved drink. The introduction of beekeeping, around the 5th century, played a significant role in its popularity. This practice is traditionally credited to Saint Modomnoc, who is said to have brought bees and, with them, the art of making mead to Ireland.

One of the most famous locations associated with mead in Ireland was the banquet hall on the Hill of Tara, known as *Tech Mid Chuarda*, which translates to "house of the circling of mead." This hall was a central place for gatherings, where mead was shared among warriors and nobles, symbolizing hospitality and unity. The drink was often infused with hazelnuts, adding a rich, nutty flavor that made it even more special.

Mead wasn't just a drink for the living; it was also deeply embedded in Irish mythology and folklore. Many legends, particularly those involving saints, mention mead as a sacred or symbolic beverage. For example, in the famous legend of the Children of Lir, mead is referenced, reflecting its importance in both everyday life and the spiritual world.

These stories highlight how mead was more than just a popular drink in medieval Ireland—it was a symbol of cultural identity, hospitality, and even spirituality. The infusion of hazelnuts and its presence in legendary tales underscore its significance, making it a drink that was cherished not just for its taste but for its place in the social and mythical fabric of Irish life.

Over time, taxes and regulations on the ingredients used in alcoholic drinks caused mead to become less common and harder to find commercially. As a result, mead faded into obscurity for a long period. However, not everyone abandoned the craft. Some monasteries, particularly in regions where grapes couldn't be grown, continued the tradition of mead-making. For these monasteries, mead was often a natural by-product of their beekeeping activities.

These religious communities kept the knowledge and practice of mead-making alive during the centuries when it was otherwise declining. In areas without vineyards, mead provided a vital source of alcohol, sustaining both monastic life and the local communities. Today, as mead enjoys a resurgence in popularity, the efforts of these monasteries remind us of the drink's deep historical roots and the role it played in preserving an important aspect of cultural heritage. The revival of mead-making today owes much to those who quietly maintained the tradition through the ages.

Mead, often called the world's oldest alcoholic beverage, has a rich and storied history that dates back thousands of years. The origins of mead are lost to time, but it is believed to have been discovered independently by various cultures around the world. The basic ingredients—honey, water, and yeast—are all naturally occurring, making mead one of the first fermented drinks humans ever crafted.

The earliest written records of mead come from ancient civilizations. In the Rigveda, a sacred text from around 1700-1100 BCE in India, a drink called *soma* is mentioned, which some scholars believe could have been mead. Similarly, the ancient Greeks wrote about a honey-based drink, with the poet Pytheas noting its use in northern lands like Thule. Mead was central to many ancient cultures, including the Celts, Saxons, and Vikings, who celebrated it in their myths, poetry, and rituals.

In medieval Europe, mead was widely consumed, especially in regions where grapes for wine were hard to cultivate. It was particularly popular in Ireland, where it became intertwined with folklore and legend. Monasteries across Europe played a crucial role in preserving the art of mead-making, as they maintained beekeeping traditions and produced mead as a by-product.

However, as wine and beer became more accessible and regulations around alcohol production grew stricter, mead gradually faded from the spotlight. For centuries, it became a rare and obscure drink, kept alive mainly in small pockets of tradition.

In recent years, mead has experienced a revival, with a growing number of craft producers and enthusiasts rediscovering its unique flavors and historical significance. Today, mead is once again enjoyed by many, connecting modern drinkers with an ancient tradition that spans the globe.

The origins of mead are ancient and somewhat mysterious, as the drink predates recorded history. It's believed that mead was independently discovered by various cultures across the globe, as its primary ingredients—honey, water, and wild yeast—are naturally occurring and likely fermented accidentally.

The earliest evidence of mead-like beverages can be traced back to around 7000 BCE in northern China, where pottery jars were found containing chemical signatures of a fermented drink made from honey, rice, and fruit. This suggests that early humans were experimenting with honey-based fermentation long before the advent of agriculture.

In Africa, there's evidence that mead was consumed in Ethiopia, where it is known as *tej*. This traditional Ethiopian honey wine is still made today and often flavored with a bitter herb called *gesho*.

In Europe, mead was deeply rooted in the cultures of ancient civilizations such as the Celts, Saxons, and Vikings. These groups celebrated mead in their myths and rituals, considering it a divine or heroic drink. The earliest written records of mead appear in the *Rigveda*, a collection of ancient Indian hymns dating back to around 1700–1100 BCE. The text mentions a drink called *soma*, which some believe may have been a type of mead.

In ancient Greece, mead was referenced by writers such as Pytheas, who noted its use in northern regions like Thule. The Greeks, like many other ancient peoples, revered mead and associated it with their gods and heroes.

As agriculture spread and viticulture became more prominent, mead began to be overshadowed by wine and beer. However, its deep roots in various cultures ensured its survival, even if only in small, traditional pockets, until its modern resurgence.

GLOBAL MEAD TRADITIONS

Mead's status as one of the oldest known alcoholic beverages is not just a historical footnote; it's a gateway into understanding human civilization, mythology, and cultural practices across millennia. This honey-based drink is much more than a simple concoction—it's a reflection of humanity's early attempts to cultivate, preserve, and celebrate life.

Mead's significance extends far beyond its role as a sustenance or celebratory drink; it has deep roots in mythology and spiritual practices. In many ancient cultures, mead was thought to possess divine properties. It wasn't merely a drink—it was considered a gift from the gods, often associated with immortality, wisdom, and life itself.

For instance, in Norse mythology, mead is directly tied to the creation of poetry and wisdom. The myth of the "mead of poetry" tells of a magical brew created from the blood of a wise being, Kvasir, which was then mixed with honey. This mead was believed to bestow the gift of eloquence and knowledge upon those who drank it, illustrating the deep cultural reverence for both the beverage and the ideals it represented. Odin, the chief of the Norse gods, was said to have stolen this mead and shared it selectively, further embedding mead in the lore of power, creativity, and divine favor.

Similarly, in ancient Greece, mead was often referred to as "ambrosia," the food and drink of the gods, conferring immortality upon them. This association with eternal life made mead a potent symbol in rituals and ceremonies, connecting the mortal with the divine.

Across cultures, mead was more than just a drink—it was a central element in social, religious, and even political practices. In medieval Europe, mead was a fixture at royal banquets and courtly gatherings, where it symbolized wealth and status. The English tradition of "wassailing," a ritual meant to ensure a good apple harvest, involved drinking mead and singing to the health of the trees, illustrating how deeply the beverage was woven into the fabric of everyday life.

In African cultures, particularly in Ethiopia, mead—known locally as tej—has been part of the social fabric for centuries. Tej is not just a drink but a symbol of hospitality, often shared during significant life events such as weddings and religious ceremonies. The preparation and sharing of tej are acts that bind communities together, reinforcing social bonds and cultural identity.

The continued relevance of mead in modern times, despite the rise of other alcoholic beverages, speaks to its enduring appeal. Today, as people seek to reconnect with natural and traditional ways of living, mead is experiencing a renaissance. Artisanal meaderies are springing up around the world, each putting its unique spin on this ancient drink. From dry, wine-like meads to sweet, spiced varieties, the modern mead movement honors the drink's rich history while exploring new possibilities.

Its origins are intertwined with the very beginnings of human society, making it a drink that connects us not only to our ancestors but to the fundamental aspects of what it means to be human. The fact that mead is still enjoyed today, in much the same way it was thousands of years ago, is a testament to its timeless appeal.

Mead is more than just an alcoholic beverage; it is a vessel for history, mythology, and culture, carrying with it the stories and traditions of countless generations. Whether sipped in a modern meadery or invoked in ancient rituals, mead continues to hold a special place in the human experience, embodying the enduring spirit of discovery, creativity, and community.

The exact origins of mead are shrouded in mystery, but archaeological evidence suggests that mead has been enjoyed for at least 4,000 years, possibly even longer. The earliest records of mead come from ancient civilizations such as the Egyptians, Greeks, and Chinese. In ancient Egypt, mead was consumed during religious ceremonies and was believed to be a gift from the gods. The Greeks, too, revered mead, calling it "ambrosia," the drink of the gods, and believed it conferred immortality.

Mead's simplicity is likely why it became so widespread; all it requires is honey, water, and yeast. This simplicity made it accessible to various cultures across the globe, from the cold climates of Scandinavia to the warm regions of Africa. Wherever bees could produce honey, mead could be made, and so it became a common thread connecting disparate societies.

In Asia, mead's history is intertwined with that of China and India. In China, mead was often consumed as part of ritual offerings to ancestors and deities. Ancient texts suggest that Chinese mead was made with various herbs and fruits, giving it a distinctive flavor. In India, mead, known as "madhu," was mentioned in the ancient Vedas, where it was celebrated as a sacred drink. Indian mead often included spices and was used in both religious rituals and medicine.

Indigenous cultures in the Americas also had their own versions of mead. The Maya of Central America, for example, brewed a honey-based alcoholic drink that was likely similar to mead, which they used in religious ceremonies. Similarly, various Native American tribes in North America made honey wines for both ceremonial and medicinal purposes.

The Revival Of Mead In Modern Times

As human societies developed, so did the craft of mead-making. The Egyptians, Greeks, and Romans all valued mead for its intoxicating effects and its association with the divine. In Ancient Egypt, mead was consumed during religious rituals and offered to gods. The Greeks believed mead to be the drink of the gods, giving it the name "Ambrosia," while the Romans used it in their ceremonies and banquets

In the past few decades, there has been a significant shift in the way people think about and consume alcohol. This change is often referred to as the "craft beverage movement." People are increasingly turning away from mass-produced, commercial drinks in favor of artisanal, small-batch products that emphasize quality, tradition, and unique flavors. This movement has paved the way for the resurgence of mead.

Craft breweries and distilleries have exploded in popularity, leading many to look beyond the usual offerings and explore other traditional drinks, like mead. Mead's rich history and diverse flavor profile make it an ideal candidate for this renewed interest in craft beverages. It's a drink that feels both ancient and fresh, offering something different from the standard options available.

One of the things that makes mead appealing today is the fact that it's made from natural ingredients. As more people become conscious of what they consume, the simplicity and purity of mead are major draws. Mead is made from honey, water, and yeast, with no need for artificial additives or preservatives. This aligns with the growing trend toward natural, organic, and locally sourced products.

Mead is incredibly versatile. It can be sweet, dry, still, sparkling, or anywhere in between. Modern mead makers are experimenting with various fruits, spices, herbs, and even hops to create a wide range of flavors. Whether you prefer a rich, dessert-like drink or something more akin to a dry wine,

CHAPTER ONE

there's likely a mead that suits your taste.

Drinking mead is like taking a sip of history. It's a drink with roots in ancient cultures around the world, from Viking warriors to African kings. For many, mead represents a connection to the past, a way to honor tradition while enjoying a timeless beverage. The resurgence of mead taps into this sense of nostalgia, offering a way to experience something our ancestors might have enjoyed. Because mead has been relatively obscure for so long, trying it offers a unique experience. For many, it's a chance to explore something new and different from the typical beer or wine, making it an intriguing option for adventurous drinkers.

The revival of mead wouldn't be possible without the rise of dedicated meaderies—businesses that specialize in producing mead. These meaderies have sprung up all over the world, each contributing to the diversity and accessibility of mead.

Modern meaderies are often run by passionate individuals who are eager to share their love of mead with others. They experiment with different types of honey, fermentation techniques, and flavor combinations, pushing the boundaries of what mead can be. Some focus on traditional recipes, while others aim to innovate, blending mead with other types of alcohol or aging it in different types of barrels for added complexity.

In addition to producing mead, these meaderies often serve as community hubs, hosting tastings, tours, and events that educate the public about mead's history and production. This helps to build a loyal customer base and spreads awareness of mead to a broader audience.

The revival of mead has also been fueled by its appearance in popular culture. Fantasy literature, movies, and TV shows, particularly those set in medieval or mythical worlds, often feature mead as the drink of choice for characters. This has introduced many people to the concept of mead, even if they've never tried it. The popularity of shows like "Game of Thrones" and "The Witcher," where mead is frequently depicted, has helped to bring the drink back into the public eye.

Moreover, the rise of craft beer festivals and renaissance fairs, where historical drinks like mead are often featured, has provided a platform for people to try mead in a fun, immersive setting. These events allow new drinkers to experience mead in a way that feels both novel and authentic.

While the revival of mead is exciting, it's not without its challenges. Mead remains a niche product compared to beer and wine, and many people are still unfamiliar with it. Educating consumers about what mead is and how it can be enjoyed is crucial for its continued growth.

Fortunately, the opportunities are vast. As mead becomes more well-known, there is potential for it to become a staple in bars and restaurants, alongside beer, wine, and cocktails. Some innovative bartenders are already creating mead-based cocktails, further expanding its appeal. Additionally, the health-conscious market, with its preference for natural and gluten-free products, offers a promising avenue for growth.

What began as an ancient, almost forgotten drink is now re-emerging as a beloved beverage for those seeking something unique, flavorful, and steeped in history. The craft beverage movement, combined with a growing appreciation for natural, artisanal products, has created the perfect environment for mead to flourish once again.

As more people discover mead and its many varieties, this honey-based drink is likely to continue its rise, finding a new place in the modern world while retaining its connection to the past. Whether enjoyed in a medieval-style tavern or a trendy urban bar, mead is poised to be a drink for the ages—once again.

2

CHAPTER TWO

UNDERSTANDING MEAD

What Is Mead?

Mead is a type of wine made from honey. Depending on how it's made, it can be either dry (with little to no sugar left) or sweet (with some sugar remaining). It can also be either still (no bubbles) or sparkling (with bubbles). You can add spices or fruit juice to it, or keep it plain.

Personally, I like to make my mead dry and still. To get it dry, I let it ferment until all the sugar is gone. To keep it still, I don't add pressure, so it doesn't become bubbly. One way to make sure mead is dry is to use less sweetener, allowing the yeast to consume all the sugar and then stop.

I often add some spices to my mead, though sometimes I leave it plain.

Mead, like other fermented drinks, has beneficial biological compounds, which can be good for those who can safely drink alcohol. However, if you find it hard to stop drinking once you start, or if you have liver issues, it's best to avoid mead altogether.

Mead can range from sweet to dry. In sweet mead, not all the sugars are consumed during fermentation, leaving a sugary taste. In dry mead, most or all of the sugar is converted into alcohol, resulting in a less sweet or even a tart flavor. Mead can also be still (with no bubbles) or sparkling (carbonated with bubbles). This depends on whether the mead is bottled under pressure or allowed to ferment in a way that captures the natural carbon dioxide. Mead can be plain, made only with honey, water, and yeast, or it can be enhanced with a variety of other ingredients. For example, spices like cinnamon or cloves can be added, or fruit juices like apple or berry can be mixed in during fermentation, creating different flavors and styles.

Ingredients Breakdown For Mead Making

However, the art of mead making extends far beyond these basics, encompassing a diverse array of ingredients that contribute to its flavor, aroma, and character. Below is an extensive breakdown of the ingredients used in mead:

Honey

Honey is the cornerstone of mead making, imbuing the beverage with its signature sweetness, complexity, and character. Its role extends beyond just a sugar source for fermentation; honey influences the flavor, aroma, color, mouthfeel, and even the aging potential of the final product.

Honey color can range from nearly transparent (water white) to dark amber or even black. Generally, lighter honeys produce more delicate, floral meads, while darker honeys yield richer, more robust flavors with notes of caramel, toffee, or molasses. The aromatic profile of honey directly translates to the mead's bouquet. Honeys with strong, distinctive aromas, like eucalyptus or lavender honey, impart those same qualities to the mead.

Types Of Honey

The choice of honey is pivotal in determining the flavor, aroma, color, and overall character of the mead. Each type of honey brings its unique qualities to the brew, influenced by the floral sources, regional climates, and processing methods

1. Wildflower Honey

Wildflower honey is made from the nectar of various flowers that bloom in a specific region and season. It's a true representation of the local flora and can vary greatly depending on location and time of year.

The flavor of wildflower honey is complex and multifaceted, often floral, fruity, or herbaceous. It can have notes ranging from light and delicate to bold and robust.

Mead made from wildflower honey often reflects the diversity of the nectar sources, resulting in a complex, layered flavor profile. The mead may have hints of different floral notes that evolve over time, making it a popular choice for both traditional and experimental meads.

2. Clover Honey

Clover honey is harvested from the nectar of clover plants, particularly common in North America and Europe. It's one of the most widely available and commonly used types of honey.

Clover honey is mild and sweet, with subtle floral undertones. It has a clean, straightforward taste that makes it versatile for various culinary applications.

Clover honey produces a light, delicate mead with a mild sweetness and a smooth finish. Its subtle flavor allows other ingredients, like fruits or spices, to shine without overpowering the mead's overall balance.

3. Orange Blossom Honey

Orange blossom honey is derived from the nectar of orange tree flowers, primarily in regions like Florida, Southern California, and Spain.
   This honey is fragrant and fruity, with distinct citrus notes that are both sweet and tangy. It has a light, golden color and a pleasant aroma reminiscent of orange groves in bloom.

Orange blossom honey is highly prized in mead-making for its bright, citrusy flavors. It imparts a lively, refreshing character to the mead, making it an excellent choice for lighter, fruit-forward meads, especially those with added citrus fruits.

4. Buckwheat Honey

Buckwheat honey comes from the nectar of buckwheat flowers, primarily grown in North America, particularly in the northern United States and Canada.
   This honey is dark, robust, and earthy, with molasses-like flavors and a strong, distinctive taste. It's less sweet than lighter honeys, with a rich, malty character and a hint of bitterness.

Mead made from buckwheat honey is full-bodied and complex, with deep, rich flavors. It's often compared to stout beer or dark molasses, making it a popular choice for robust, hearty meads, particularly those aged in oak barrels or flavored with spices.

5. Manuka Honey

Manuka honey is produced from the nectar of the Manuka tree (Leptospermum scoparium), native to New Zealand and parts of Australia. It's renowned for its medicinal properties and is often used for therapeutic purposes.

Manuka honey has a strong, earthy flavor with herbal and slightly bitter notes. It's less sweet than many other honeys and has a thick, viscous texture.

Mead made from Manuka honey tends to be bold and complex, with a unique, slightly medicinal taste. It's often used in meads intended for health benefits or those looking for a distinctive, unconventional flavor profile. The strong flavor of Manuka honey can dominate the mead, so it's typically used sparingly or blended with other honeys.

6. Acacia Honey

Acacia honey is derived from the nectar of the black locust tree (Robinia pseudoacacia), commonly known as the acacia tree, and is mainly produced in Europe and North America. In

Acacia honey is light, delicate, and mildly sweet, with subtle floral undertones and a hint of vanilla. It's known for its clarity and light golden color.

Acacia honey produces a mead that is soft and smooth, with a gentle sweetness and a clean finish. Its mild flavor makes it an excellent base for meads where subtlety is desired, allowing other ingredients like herbs or light fruits to come through without being overshadowed.

7. Heather Honey

Heather honey is harvested from the nectar of heather plants, particularly in moorlands and heathlands across the United Kingdom, Ireland, and parts of Europe.

Heather honey is strong and aromatic, with a slightly bitter, tangy taste and a gelatinous texture. It has a distinctive, smoky flavor with floral and herbal notes.

Heather honey mead is typically full-bodied and complex, with a unique flavor profile that can include smoky, earthy, and slightly bitter elements. It's often used in traditional meads and those intended for long aging, as the bold flavors can mellow and develop over time.

8. Sage Honey

Sage honey is primarily produced in the western United States, particularly California, from the nectar of sage plants.

Sage honey is light and mild, with a subtle herbal flavor and a smooth, buttery texture. It's known for its clarity and pale color.

Sage honey is ideal for creating a delicate and balanced mead, where its mild flavor allows the honey's natural sweetness to shine without being overpowering. It's often used in traditional meads and those where the honey flavor is intended to be the star.

9. Eucalyptus Honey

Eucalyptus honey is harvested from the nectar of eucalyptus trees, found in Australia, South Africa, and parts of the United States.

This honey has a strong, distinct flavor with medicinal, menthol-like notes. It's less sweet than other honeys, with a slightly smoky and herbal taste.

Eucalyptus honey imparts a unique, refreshing flavor to mead, often with a cooling aftertaste. It's particularly well-suited for spiced or herbal meads, where its distinctive character can complement other bold flavors.

10. Tupelo Honey

Tupelo honey is produced from the nectar of the white tupelo tree (Nyssa ogeche), found in the southeastern United States, particularly in Florida and Georgia.

Tupelo honey is prized for its buttery, smooth taste with a light, fruity flavor and a hint of floral notes. It has a high fructose content, which means it resists crystallization.

Tupelo honey is often considered the gold standard for mead-making, producing a luxurious, smooth mead with a refined sweetness and a long, clean finish. Its complex flavor profile makes it a favorite among mead makers for both traditional and showcase meads.

11. Lavender Honey

This honey is made from the nectar of lavender flowers, primarily in regions like Provence in France, where lavender fields are abundant.

It is aromatic and fragrant, with a pronounced floral taste and a slightly herbal undertone. It has a sweet, delicate flavor with a hint of spiciness.

Lavender honey is ideal for meads where floral notes are desired. It imparts a distinct, aromatic flavor that can be enhanced with additional lavender or other complementary herbs, resulting in a mead that is both fragrant and flavorful.

12. Blackberry Honey

Blackberry honey is derived from the nectar of blackberry blossoms, found in regions with abundant wild or cultivated blackberry plants.

Blackberry honey is fruity and sweet, with a rich, berry-like flavor and a dark amber color. It has a robust, tangy taste that's less floral and more fruity.

Mead made from blackberry honey has a rich, fruity character, often with a deep amber hue. It's particularly well-suited for melomels (fruit meads) and other meads where a strong, fruity flavor is desired.

13. Fireweed Honey

Fireweed honey comes from the nectar of fireweed plants, which thrive in areas affected by forest fires or cleared land, particularly in the Pacific Northwest and Alaska.

Fireweed honey is light, sweet, and floral, with a delicate taste and a pale, almost transparent color. It has a slightly spicy, tea-like flavor.

Fireweed honey produces a light, delicate mead with a subtle floral character. It's often used in traditional meads where a light, refreshing flavor is desired, allowing the honey's unique characteristics to shine.

Always remember that the choice of honey greatly influences the mead's flavor. For instance, darker honeys like buckwheat or heather provide stronger, more pronounced flavors, while lighter honeys like acacia or clover are subtler.

The amount of sugar in honey directly affects the fermentation process and alcohol content. Higher sugar content can lead to higher alcohol content or a sweeter mead if fermentation is stopped early.

Honey naturally has a low pH, contributing to the overall acidity of the mead, which affects its preservation and taste.

Water

Water is often the unsung hero in mead-making. While honey gets the spotlight, the quality and characteristics of the water used can significantly influence the final product. Understanding the role of water in mead-making, and how to optimize it, is crucial for crafting a delicious and well-balanced mead.

Water is the primary solvent in mead, dissolving honey, nutrients, and any additional ingredients like fruits or spices. It makes up the majority of the mead's volume, which means its quality directly affects the taste, mouthfeel, and overall stability of the final product. Essentially, the water serves as the canvas on which the flavors of the honey and other ingredients are painted.

Water Purity

Importance of Clean Water

The purity of the water used in mead-making is paramount. Impurities such as chlorine, chloramine, heavy metals, or organic matter can interfere with fermentation, potentially leading to off-flavors or even stalled fermentation. Clean, fresh water ensures that the yeast can thrive, and that the subtle flavors of the honey are not masked or distorted.

Filtration

Most mead makers use filtered water to remove chlorine and other impurities. Simple carbon filters, such as those used in household water pitchers, can effectively remove chlorine and chloramine, making the water safer and better suited for fermentation.

Bottled Water

Some mead makers prefer to use bottled spring water or distilled water to ensure purity. However, it's important to note that distilled water lacks minerals, which may be beneficial for yeast health.

The mineral content of water, particularly calcium and magnesium, determines whether water is considered hard or soft.

- **Hard Water:** Contains higher levels of calcium and magnesium, which can enhance yeast activity, improve fermentation, and contribute to a fuller mouthfeel in the final mead. However, too much hardness can lead to bitterness or astringency, so balance is key.
- **Soft Water:** Lacks these minerals, resulting in a cleaner, more neutral taste. While it might produce a smoother mead, it can also lead to slower fermentation if the yeast doesn't receive enough nutrients.

For mead-making, moderately hard water is often preferred. It provides the necessary minerals to support healthy yeast growth and a well-rounded flavor without overwhelming the delicate honey notes.

Water with a higher pH (more alkaline) can result in a flatter-tasting mead, as it might neutralize some of the natural acidity from the honey and other ingredients. Conversely, water that is too acidic can lead to a sharp, harsh taste in the final product.

Water Sources

Tap Water

Most mead makers use tap water, as long as it's treated to remove chlorine and chloramine. However, the mineral content and pH can vary widely depending on the local water supply.

Well Water

Well water can be an excellent choice, particularly if it's naturally clean and has a balanced mineral content. However, it should be tested for contaminants, as well water can sometimes contain high levels of iron, sulfur, or other minerals that could negatively affect the mead.

Spring Water

Bottled spring water is often used for its purity and balanced mineral content. It's a safe choice, particularly if your tap water is heavily treated or has undesirable characteristics.

Distilled Water

Distilled water is completely free of minerals, making it a very neutral choice. However, because it lacks the minerals necessary for yeast health, mead makers using distilled water often add nutrients or mineral salts to compensate.

The ratio of honey to water is crucial in determining the sweetness and alcohol content of the mead. A higher honey-to-water ratio will result in a sweeter, more alcoholic mead, while a lower ratio will produce a drier, lighter mead.

When adding water to dilute the honey, care should be taken to ensure that the dilution doesn't strip away too much of the honey's flavor. The water should enhance the honey's characteristics, not overshadow them.

When mixing honey with water, it's common to use warm water to help dissolve the honey more easily. However, the water shouldn't be too hot, as high temperatures can degrade the honey's delicate flavors and aromas.

The temperature of the water (and the overall must) during fermentation is critical. Yeast prefers a stable temperature, usually between 65-75°F (18-24°C). Water that's too cold can slow down fermentation, while water that's too warm can lead to off-flavors and a harsh, fusel alcohol taste.

When making meads with fruits, spices, or herbs, the water's role becomes even more important. It acts as the medium that extracts and carries the flavors from these ingredients throughout the mead. The water should be of high enough quality to complement, rather than compete with, these additional flavors.

If you back-sweeten or carbonate your mead, the water used in these processes should match the original water as closely as possible to maintain consistency in flavor and mouthfeel.

Yeast

Yeast is one of the most crucial ingredients in mead-making, as it is responsible for the fermentation process that turns honey and water into alcohol.

Yeast are microscopic fungi that consume sugars—primarily glucose, fructose, and sucrose in honey—during fermentation. They convert these sugars into alcohol (ethanol) and carbon dioxide (CO_2). This process is what creates the alcohol content in mead.

Yeast contributes significantly to the flavor profile of mead. Different strains produce varying levels of esters, phenols, and other compounds that can add fruity, spicy, or floral notes to the final product.

Types Of Yeast

Wine Yeast

Commonly used for mead due to its ability to handle higher alcohol levels. Popular strains include:

- **Lalvin EC-1118**: Known for its clean fermentation and high alcohol tolerance (up to 18%). It's a reliable choice for most meads, especially those where you want the honey's flavor to shine without too much interference from the yeast.
- **Lalvin D-47**: Produces a smooth, full-bodied mead with more complex

flavors. It is ideal for meads where you want to enhance the mouthfeel and introduce subtle notes of citrus and honey.
- **K1V-1116**: Known for producing fruity and floral esters, this yeast is ideal for lighter, aromatic meads.

Beer Yeast

Sometimes used for specific styles like braggot (a hybrid of beer and mead). These strains may not tolerate as high alcohol content as wine yeast but can add interesting malt-like flavors.

Wild Yeast

Some mead makers use wild yeast for a more natural and unpredictable fermentation. This can produce unique flavors but comes with the risk of spoilage or off-flavors due to unwanted microorganisms.

Honey lacks many of the nutrients that yeast need to thrive, such as nitrogen, amino acids, vitamins, and minerals. Without these, yeast may struggle to complete fermentation, leading to stuck fermentations or off-flavors.

Proper yeast rehydration is crucial for ensuring a healthy fermentation. Yeast should be rehydrated in water at around 100-105°F (37-40°C) for about 15 minutes before being added to the must (the honey-water mixture).

The amount of yeast added to the must is important. Under-pitching can stress the yeast, leading to incomplete fermentation and off-flavors, while over-pitching can result in a very rapid fermentation that may produce unwanted by-products.

Yeast is sensitive to temperature. Most wine yeasts perform best in the range of 60-75°F (15-24°C). Higher temperatures can cause yeast to produce fusel alcohols, which impart harsh, solvent-like flavors.

Yeast by-Products and Their Effects

- **Esters**: These are fruity compounds that can add complexity to the mead. The type and amount of esters depend on the yeast strain and fermentation conditions.
- **Phenols**: Some yeasts produce phenolic compounds that can add spicy or medicinal flavors. These are more common in beer yeasts and certain wild yeast strains.
- **Fusel Alcohols**: Higher alcohols that are produced at warmer fermentation temperatures or by stressed yeast. These can give the mead a harsh taste and should be minimized.

A stuck fermentation occurs when yeast become dormant or die before consuming all the sugars. This can be caused by nutrient deficiencies, excessive alcohol levels, or poor yeast health.

Restarting fermentation can involve adding more yeast, particularly a tolerant strain like EC-1118, and ensuring adequate nutrients and oxygenation.

After fermentation, dead yeast cells (lees) settle at the bottom. Over time, these cells break down, releasing flavors into the mead. This process, known as autolysis, can add complexity and body to the mead if managed carefully.

To avoid off-flavors from extended contact with the lees, mead makers often rack (transfer) the mead off the lees after primary fermentation is complete.

With the growing popularity of mead, yeast labs are developing strains specifically tailored to mead-making. These may offer unique flavor profiles or be engineered to perform better in the high-sugar, low-nutrient environment of mead must.

Some mead makers experiment with blending different yeast strains to create complex flavor profiles or to capitalize on the strengths of each strain.

Finding the perfect yeast for your mead may require experimentation. Different batches might benefit from different strains, so keeping detailed records of your yeast choices and their results is crucial.

Yeast is not just a tool for alcohol production in mead-making; it is an active participant in crafting the final flavor, aroma, and overall experience of the mead.

Additional Fermentables (Fruits,Spices,Syrups)

When we talk about additional fermentables in mead-making, we're referring to any sugars or sugar-containing ingredients added to the honey and water mixture (must) to influence the flavor, aroma, color, or alcohol content of the mead. These ingredients can be as varied as fruits, grains, herbs, or even other types of sugars

1. Fruits (Melomels)

Common fruits used in mead include berries (raspberries, blueberries, blackberries), stone fruits (peaches, plums, cherries), apples, pears, citrus fruits, and tropical fruits (mango, pineapple). Each type of fruit imparts its unique flavor, color, and aroma to the mead.

Fruits can be added during primary fermentation, where the yeast will ferment the sugars in the fruit along with the honey, or during secondary fermentation, where the fruit's flavors are more likely to be preserved and pronounced.

Fresh, frozen, dried, or pureed fruits can all be used. Fresh fruits provide the most natural flavors but can be seasonal, while frozen fruits often offer better availability and can help break down cell walls, releasing more juice. Dried fruits like raisins or apricots can concentrate flavors and add complexity.

Fruits add additional sugars, which can increase the alcohol content of the mead. However, they also introduce pectin, which may require the addition of pectic enzymes to prevent haze in the finished product. The acids and tannins in fruit can also affect the pH balance, fermentation rate, and overall mouthfeel of the mead.

2. Spices and Herbs (Metheglins)

Cinnamon, cloves, ginger, vanilla, cardamom, and herbs like mint, thyme, or rosemary are commonly used to create a metheglin (spiced mead). Each spice or herb brings its unique character, which can range from warming and earthy to fresh and aromatic.

Spices and herbs can be added during primary or secondary fermentation, or even during aging. Adding them earlier allows the yeast to interact with these compounds, which can soften and integrate their flavors, while later additions result in more intense and defined spice notes.

The intensity of these flavors can be controlled by the quantity used and the duration they remain in contact with the must. Over-spicing can lead to overpowering flavors, so it's often recommended to start with a small amount and adjust as needed.

3. Grains and Malt (Braggots)

Braggots are a hybrid of mead and beer, where malted grains are added to the must. The malt contributes fermentable sugars, proteins, and flavors typical of beer, such as caramel, toast, or chocolate notes.

Barley is the most common grain used, but other grains like wheat, rye, or oats can be added for different textures and flavors.

The grains must be mashed (steeped in hot water) to convert their starches into fermentable sugars before being added to the honey-water mixture. This process is similar to brewing beer and requires additional equipment and steps.

The key challenge in braggots is balancing the malt's rich, often heavy flavors with the delicate sweetness of honey. Careful selection of honey and grain types, as well as the yeast strain, can help achieve this balance.

4. Sugars and Syrups

Cane Sugar, Brown Sugar, or Molasses

These can be added to increase the alcohol content and introduce additional flavors. Brown sugar and molasses, for instance, add rich, caramelized notes, but too much can overpower the honey's subtleties.

Maple Syrup or Agave Nectar

These alternative sugars offer unique flavors. Maple syrup can impart a woody, smoky sweetness, while agave nectar adds a lighter, floral quality. These sugars are often used in small quantities due to their strong flavors.

Invert Sugar

Invert sugar, which is sugar broken down into glucose and fructose, can be used to boost alcohol content without affecting the honey's flavor too much. It's more fermentable than regular sugar and can help create a smoother mead.

5. Fruit Juices

Instead of whole fruits, fruit juices (like apple, cranberry, or grape juice) can be added to the must. Juices can simplify the process by avoiding the mess of dealing with whole fruits and are often pasteurized, reducing the risk of contamination.

Fruit concentrates are another option, providing a more intense flavor without adding as much liquid volume. They are especially useful when aiming for a specific flavor profile or when fresh fruits are out of season.

When using fruit juices or concentrates, it's important to account for the added acidity and sugar content, as this can impact the fermentation process and final taste of the mead.

6. Honey Varietals

Although honey is the primary fermentable in mead, using multiple varietals can add complexity. For example, blending clover honey with a more robust buckwheat honey can create a mead with layers of flavor.

Some mead makers add additional honey during secondary fermentation (a process called "back-sweetening") to sweeten the mead or to accentuate honey flavors after primary fermentation has subdued them.

7. Unconventional Fermentables

Carrots, beets, and even sweet potatoes have been used in mead-making to add earthy sweetness and unique color. These are more experimental but can produce surprisingly delicious results.

Flowers like hibiscus, elderflower, or lavender can be added to the must for their delicate flavors and aromas. They usually contribute little in the way of fermentable sugars but can profoundly influence the mead's bouquet and taste.

8. Fermentation Management with Additional Fermentables

When additional fermentables are used, it's essential to monitor the sugar content (measured as Specific Gravity or Brix) throughout fermentation. This ensures that fermentation proceeds as expected and allows adjustments if the yeast struggles with the extra sugars.

Adding fruits or other fermentables may change the nutrient profile of the must, requiring adjustments in yeast nutrient additions to avoid stuck fermentation or off-flavors.

When bottling mead that contains additional fermentables, especially if back-sweetened or with added fruit, there is a risk of re-fermentation if not properly stabilized. This can lead to carbonation in the bottle or even bottle bombs.

Clarifying Agents

Clarifying agents, also known as fining agents, are substances added to mead (and other fermented beverages like wine and beer) to help remove suspended particles that cause haze or cloudiness. These agents work by binding to the unwanted particles, making them heavier so they can settle to the bottom, where they can be easily separated from the clear mead.

Clear mead is often more visually appealing than cloudy mead, especially when bottled in transparent glass. Clarity is a hallmark of quality in many beverage categories.

Removing suspended particles can also improve the stability of the mead, reducing the likelihood of spoilage or off-flavors developing over time.

Some suspended particles can impart harsh or undesirable flavors to the mead. Clarifying agents help remove these, leading to a cleaner taste and smoother mouthfeel.

Types of Clarifying Agents

Bentonite

Bentonite is a type of clay made up of volcanic ash. It's negatively charged, which allows it to attract and bind positively charged particles like proteins and tannins.

Bentonite is typically added early in the fermentation process. It's hydrated in water before being mixed into the mead. As it settles, it drags down particles that cause haze.

Bentonite is highly effective at removing proteins and is one of the most commonly used fining agents in mead and wine-making. It is also relatively inexpensive.

Overuse of bentonite can strip away desirable flavors or aromas. Additionally, it works best at higher temperatures (around 85°F or 29°C), which may not be ideal for all fermentation schedules.

Sparkolloid

Sparkolloid is a blend of polysaccharides (complex carbohydrates) derived from plants. It works through both electrostatic attraction and physical adsorption.

It is typically dissolved in hot water and then added to the mead. It's usually used towards the end of fermentation, as it's particularly effective at removing yeast cells and other fine particulates.

Sparkolloid can produce a brilliantly clear mead with minimal impact on flavor. It's also effective across a wide range of temperatures.

It's important to handle Sparkolloid carefully, as it can become gummy if not prepared correctly. Some users also report that it can be difficult to fully remove from the mead.

Isinglass

Isinglass is a fining agent derived from the swim bladders of fish, primarily sturgeon. It's a form of collagen, like gelatin, and is primarily used in wine and beer.

Isinglass is typically added during secondary fermentation or in the aging process. It works best in cooler temperatures and is often used to polish a mead that's already fairly clear.

Isinglass is gentle on the mead's flavor and aroma and can produce a very clear final product. It's also effective at removing yeast and some types of haze. Like gelatin, isinglass is animal-based, which can be an issue for vegan mead makers. It's also temperature-sensitive and may require cold stabilization to work effectively.

Polyvinylpolypyrrolidone (PVPP)

PVPP is a synthetic polymer that acts as a fining agent by binding to polyphenols and tannins, which can cause haze and astringency.

PVPP adsorbs polyphenols and tannins, causing them to precipitate out of the mead. It is especially useful for reducing bitterness and astringency in meads that have been over-exposed to tannic fruits or wood.

PVPP is usually added after fermentation, during the aging process. It is mixed with water and added to the mead, where it works relatively quickly.

PVPP is effective at improving both clarity and flavor, particularly in meads that have an overly tannic profile. It's a good choice for meads that need refinement in both taste and appearance.

Natural Clarification Methods

Cold Crashing

Cold crashing involves lowering the temperature of the mead to near freezing after fermentation is complete. The cold temperature causes yeast and other particles to flocculate (clump together) and settle out of the mead.

The mead is typically placed in a refrigerator or cold storage for several days to weeks. After cold crashing, the mead is carefully racked off the sediment.

Cold crashing is a natural and gentle way to clarify mead, particularly effective at removing yeast. It doesn't involve any additives, making it a good option for those who prefer minimal intervention. However, it may not be sufficient for more stubborn hazes.

Racking

Racking involves transferring mead from one container to another, leaving the sediment (lees) behind. This process can help to gradually clarify the mead over time. Racking is usually done multiple times during the aging process, particularly after primary fermentation and before bottling.

While racking alone may not result in crystal-clear mead, it is an essential part of the clarification process. It allows for the gradual removal of sediment and can be used in conjunction with other clarifying methods.

3

CHAPTER THREE

TYPES OF MEAD

Traditional Mead

Traditional mead is the foundational form of mead, consisting only of honey, water, and yeast. This simplicity allows the natural flavors of the honey to take center stage. The type of honey used is crucial in traditional mead because its flavor profile directly influences the final product. For example, a traditional mead made with clover honey might be light and floral, while one made with buckwheat honey could be dark and earthy with a molasses-like richness.

The fermentation process in traditional mead is relatively straightforward. Yeast consumes the sugars in the honey, producing alcohol and carbon dioxide. The amount of honey used and the length of fermentation determine the mead's sweetness and alcohol content. A shorter fermentation with more honey will result in a sweeter mead, while a longer fermentation with less honey will produce a drier, more alcoholic beverage.

Traditional meads are often aged for several months or even years to allow the flavors to mellow and develop. Aging can bring out subtle notes in the honey that might not be immediately apparent in younger meads. The beauty of traditional mead lies in its simplicity—there are no added fruits, spices, or herbs to distract from the honey's natural essence.

Melomel

Melomel is a category of mead that includes any mead made with fruits. The fruit can be added during or after fermentation, and it adds its own sugars, flavors, and colors to the mead. The type of fruit used in melomel can significantly change the character of the drink.

Cyser is a popular type of melomel made with apples or apple juice. It's essentially a marriage between mead and hard cider. The apple brings a crisp, refreshing quality to the mead, while the honey adds a layer of sweetness and complexity. Cyser can be made from various apple varieties, ranging from tart and sharp to sweet and mellow, giving it a wide range of possible flavor profiles.

Pyment is another well-known melomel made with grapes or grape juice. It's a blend of mead and wine, with the grapes' tannins and acidity complementing the honey's sweetness. Depending on the grape variety, pyment can resemble a light, fruity wine or a bold, full-bodied red. The honey in pyment often softens the sharpness of the wine, creating a balanced, smooth drink.

Berry Melomels are made with berries like raspberries, blackberries, or blueberries. These fruits add a tart, fruity character to the mead, often resulting in a bright, vibrant drink with a beautiful color. The natural acidity of the berries balances the honey's sweetness, creating a refreshing beverage that can be enjoyed young or aged for more complexity.

Melomels are incredibly versatile, allowing mead makers to experiment with different fruit combinations and fermentation techniques to create unique flavors. The fruit adds a layer of complexity to the mead, making it a favorite among those who enjoy more robust, flavorful drinks.

Metheglin

Metheglin is a type of mead flavored with spices and herbs. The word "metheglin" comes from the Welsh word "meddyglyn," meaning medicinal liquor, as these spiced meads were historically used for their supposed health benefits. Today, metheglin is enjoyed for its rich, complex flavors.

Spices like cinnamon, cloves, nutmeg, and ginger are commonly used in metheglin. These spices can give the mead a warming, comforting character, making it perfect for sipping by the fire on a cold winter evening. The level of spicing can vary from a subtle hint to a bold, spicy kick, depending on the recipe.

Herbs such as mint, chamomile, and lavender can also be used to flavor metheglin. These herbs add a floral or herbal note to the mead, creating a delicate, aromatic drink. Lavender metheglin, for example, has a soft, soothing flavor with a pleasant floral aroma that pairs well with honey's natural sweetness.

The beauty of metheglin lies in its versatility. Almost any spice or herb can be used to flavor the mead, allowing for endless experimentation. The spices and herbs used in metheglin can also help to preserve the mead and enhance its shelf life.

Hippocras

Hippocras is a subtype of pyment, a spiced grape mead with a long history. Named after the ancient physician Hippocrates, hippocras was traditionally used for medicinal purposes. It was popular in medieval Europe, often served at banquets and feasts as a digestif.

The spices used in hippocras are similar to those found in metheglin, such as cinnamon, ginger, and cloves. These spices were prized in medieval times for their exotic origins and were often used to show off wealth and sophistication. The combination of grapes and spices creates a rich, complex drink that is both sweet and warming.

Hippocras is typically sweeter than other meads due to the addition of sugar or honey to balance the spices' strong flavors. It's often enjoyed as a dessert wine, served chilled or at room temperature. The spices give hippocras a festive character, making it a popular choice for holiday celebrations.

Bochet

Bochet is a unique type of mead made by caramelizing the honey before fermentation. This process involves heating the honey until it turns a deep amber color and develops rich, toasty flavors. The result is a mead with a distinct caramel, toffee, and burnt sugar character that sets it apart from other meads.

Caramelizing the honey can be tricky, as it requires careful control of the heat to avoid burning the honey. However, the effort is worth it for the complex flavors it imparts to the mead. Bochet is often darker in color than traditional mead, with a deep, rich flavor that pairs well with desserts or as a stand-alone sipping drink.

Bochet can be made dry or sweet, depending on the amount of honey used and the length of fermentation. It's also a great base for aging, as the caramelized flavors develop and mature over time. A well-aged bochet can have a smooth, velvety texture with layers of complex flavors that unfold with each sip.

Braggot

Braggot is a hybrid beverage made by combining malted barley or other grains with honey. It's essentially a cross between beer and mead, with the malt providing the body and hoppy bitterness of beer, while the honey adds sweetness and complexity. Braggot has a long history, dating back to medieval times when it was enjoyed by both peasants and nobility alike.

The ratio of malt to honey in braggot can vary widely, resulting in a wide range of flavors and styles. Some braggots are more beer-like, with a strong malt and hop presence, while others lean more towards mead, with the honey taking center stage. The type of malt and hops used also influences the final flavor, with darker malts adding a roasted, chocolatey note and hops contributing bitterness and aroma.

Braggot is a great entry point for beer lovers interested in mead, as it combines familiar beer flavors with the unique characteristics of honey. It can be carbonated like beer or still like mead, and it can be aged to develop more complex flavors. Whether you prefer a hoppy, bitter drink or a smooth, sweet beverage, there's a braggot out there to suit your taste.

Hydromel

Hydromel is a lighter, lower-alcohol version of mead, typically with an alcohol content below 7%. The word "hydromel" comes from the Greek words "hydro," meaning water, and "meli," meaning honey. In essence, hydromel is a watered-down mead that's light, refreshing, and easy to drink.

Hydromel is often sparkling, with a gentle effervescence that makes it a perfect summer drink. It's typically less sweet than other meads, with a crisp, clean finish. The lower alcohol content makes hydromel more sessionable, meaning you can enjoy a few glasses without feeling too heavy.

In France, the term "hydromel" is used to describe any mead, but outside of France, it specifically refers to this lighter style. Hydromel is often enjoyed young, without the need for long aging. Its light, refreshing character makes it a popular choice for casual drinking, picnics, or as a palate cleanser between courses at a meal.

Sack Mead

Sack mead is a term used to describe a very sweet, strong mead with a higher alcohol content. The name "sack" comes from the old English term for fortified wine. Sack mead is made by using a large amount of honey relative to water, resulting in a rich, sweet beverage with a high residual sugar content.

Sack mead can be thought of as the mead equivalent of a dessert wine. It's often enjoyed in small quantities due to its sweetness and potency. The high sugar content means that not all the honey ferments into alcohol, leaving behind a rich, syrupy texture.

Sack mead is ideal for aging, as the high sugar and alcohol content act as preservatives, allowing the flavors to develop and mature over time. A well-aged sack mead can have a smooth, velvety texture with layers of complex flavors, such as caramel, vanilla, and dried fruit.

Sparkling Mead

Sparkling mead is the effervescent cousin of traditional mead, infused with bubbles that transform the drinking experience. Just as champagne is to wine, sparkling mead is to still mead—a celebratory, lively version that's perfect for special occasions or when you just want something a bit more festive.

The bubbles in sparkling mead can be achieved in a couple of ways. The traditional method, similar to how champagne is made, involves a secondary fermentation in the bottle. After the initial fermentation, a small amount of sugar or honey is added to the mead before it's bottled. This additional sugar feeds the remaining yeast, which produces carbon dioxide that gets trapped in the bottle, creating the bubbles. This method is known as "méthode champenoise" or the traditional method.

Another way to create sparkling mead is through force carbonation, where carbon dioxide is directly injected into the mead, similar to how soda is carbonated. This method is quicker and allows for more control over the level of carbonation, but it doesn't impart the same depth of flavor as the traditional method.

Sparkling mead can be made from any base mead—traditional, melomel, metheglin, or others. The carbonation adds a crispness and a refreshing quality that can make even the sweetest meads feel lighter and more drinkable. The bubbles also enhance the aroma, bringing the honey's floral notes to the forefront.

Because of its refreshing nature, sparkling mead is often enjoyed chilled and is a popular choice for warm-weather gatherings. It pairs well with a wide range of foods, from appetizers like cheese and fruit to lighter main courses like salads and seafood. Its versatility and celebratory vibe make sparkling mead a fun and approachable option for those new to mead.

Show Mead

Show mead is the ultimate showcase of honey's natural flavor, where the goal is to highlight the quality of the honey without any additional ingredients altering its character. Unlike other types of mead that might include fruits, spices, or herbs, show mead is strictly made with just honey, water, and yeast. This minimalist approach puts the honey's unique flavor profile at the center of attention.

The key to making a great show mead is using high-quality honey, as it's the star of the show. Different types of honey can result in vastly different meads. For example, a show mead made with orange blossom honey will have a citrusy, floral aroma, while one made with wildflower honey might have a more complex, earthy character.

Because show mead doesn't rely on additional flavorings, the fermentation process needs to be carefully controlled to ensure that the honey's delicate flavors aren't overshadowed by off-flavors from the yeast. This often involves using specific strains of yeast that are known for their clean fermentation characteristics and paying close attention to temperature control during fermentation.

Show mead can be dry, semi-sweet, or sweet, depending on the amount of honey used and the length of fermentation. The clarity and color of the mead are also important, as a clear, golden mead is more visually appealing and reflects the purity of the ingredients.

For those interested in exploring the nuances of different honey varieties, show mead offers a pure and unadulterated experience. It's often aged for several months to allow the flavors to fully develop, resulting in a smooth, well-rounded drink that truly captures the essence of the honey used.

Session Mead

Session mead is a modern term that describes a lower-alcohol, more approachable version of mead, typically with an alcohol content between 3% and 7%. The concept of a "session" beverage comes from the idea of a drink that's easy to enjoy over an extended period—something you can sip on throughout a social gathering without feeling overly intoxicated. Session mead fits this bill perfectly, offering all the flavors of traditional mead in a lighter, more refreshing package.

Session mead is often carbonated, adding to its light and refreshing nature. The carbonation, combined with a lower alcohol content, makes it a great option for warm weather or outdoor events where you might want something crisp and thirst-quenching. Unlike traditional meads, which can be quite rich and heavy, session meads are designed to be more drinkable, with a balanced sweetness and acidity that keeps you coming back for more.

The lower alcohol content in session mead is achieved by using less honey or fermenting for a shorter period. This means that session meads are often less sweet than their higher-alcohol counterparts, with a cleaner, crisper finish. The flavor profile can range from fruity and floral to dry and slightly tart, depending on the type of honey and any additional ingredients used.

Session mead is a great introduction to the world of mead for beginners, as it's less intense and more versatile than traditional mead. It pairs well with a variety of foods, from light appetizers to grilled meats, and is a popular choice for those looking for a more casual, everyday mead.

II

THE MEAD MAKING PROCESS

4

CHAPTER FOUR

GETTING STARTED

Essential Equipment and Tools

To make mead, having the right tools and equipment is essential for a smooth process and to ensure high-quality results. We will be talking about the most essential items you'll need, from basic to more advanced tools, all of which contributes to a successful mead production.

1. Fermentation Vessel

This is where your honey, water, yeast, and any other ingredients (such as fruits or spices) will ferment into mead.

Types of fermentation vessel

- **Carboys**: These glass or plastic jugs (usually 1-5 gallons) are preferred by many mead makers. They come in clear or colored varieties, with darker ones often preferred to limit light exposure.

- **Fermentation Buckets**: These food-grade plastic buckets are usually around 5 gallons in size. They are easy to clean and great for larger batches.

The fermentation vessel is where your mead comes to life. It needs to be non-reactive (no metal) and provide enough space for the mead to ferment while preventing contamination.

2. Airlock and Bung

The airlock allows gases to escape during fermentation while keeping outside air and contaminants out. It fits into a rubber bung or stopper that seals the fermentation vessel.

Types of Airlock and Bung

- **S-shaped airlocks**: Traditional design that bubbles visibly, letting you know fermentation is happening.
- **3-piece airlocks**: Easier to clean and handle, though both types work equally well.

Airlocks prevent the risk of oxidation and contamination while maintaining a sanitary environment for fermentation. They also offer a visual indicator of the fermentation process.

3. Hydrometer

A hydrometer measures the density of a liquid relative to water. In mead-making, it's used to measure sugar content before, during, and after fermentation.

Tracking the specific gravity (SG) helps you to:

- Know the potential alcohol content of your mead.
- Determine when fermentation is complete.
- Avoid bottling too early (which could cause bottle explosions).

Tip: A **test jar** is used in combination with the hydrometer to collect samples of your mead for accurate readings.

4. Sanitizing Solution

A sanitizing solution like StarSan or Iodophor is crucial for cleaning your equipment to prevent unwanted bacteria, yeast, or mold from spoiling your mead.

Sanitation is key in any fermentation process. Even the smallest contamination can ruin a batch, giving your mead off-flavors or causing infections.

5. Stirring Spoon or Paddle

A long-handled, non-reactive spoon (usually plastic or stainless steel) or paddle is necessary for mixing ingredients.

You'll need to stir honey into water for even dissolution and occasionally stir in nutrients or other additives during fermentation.

6. Thermometer

A simple kitchen thermometer or a brewing-specific thermometer is used to monitor the temperature of your must (the mixture of honey, water, and other ingredients).

Temperature control is critical in mead making. Too high a temperature can kill yeast, while too low can slow fermentation significantly.

7. Siphon and Tubing

A racking siphon and food-grade tubing are used to transfer liquid from one container to another while leaving sediment behind.

Siphoning ensures that you leave behind the yeast sediment (also known as "lees") that settles at the bottom of the fermentation vessel. This is crucial to improve clarity and taste, as sediment can impart off-flavors if left too long.

8. Bottles and Corks

Once fermentation is complete, your mead will need to be bottled. Bottles come in various shapes and sizes, with cork or swing-top closures.

Types of bottles and Corks

- **Wine Bottles**: For still meads, standard wine bottles with cork closures are common.
- **Beer Bottles**: For sparkling meads, swing-top bottles or crown-capped beer bottles work well.

Bottles are the final storage containers for your mead. Corks or caps help preserve freshness and aging potential.

9. Bottle Filler and Capper/Corker

- **Bottle Filler**: A handy tool that attaches to your siphon, making the process of filling bottles easier and more precise.
- **Capper/Corker**: A device that helps you securely seal bottles. Cappers are used for crown-capped bottles, and corkers are for wine bottles with corks.

10. PH Meter or Strips

A pH meter or pH test strips help you measure the acidity of your must.
Yeast thrives in specific pH ranges (around 3.7 to 4.5). By monitoring pH,

you can adjust your must to create an environment where yeast can work efficiently.

11. Yeast Nutrient and Energizer

Yeast nutrients and energizers provide essential nitrogen, vitamins, and minerals that help yeast ferment honey efficiently.

Honey alone lacks many of the nutrients yeast needs to ferment efficiently. Adding nutrients ensures healthy fermentation, reduces lag time, and minimizes off-flavors.

12. Refractometer (Optional but useful)

A refractometer measures the sugar content of your must before fermentation using only a small drop of liquid. It works by measuring how light bends (refracts) through the liquid.

It offers a quick and precise way to measure sugar levels without needing large samples. However, after fermentation begins, corrections are needed for alcohol content.

13. Degassing Wand or Whip (Optional)

A degassing wand or whip, usually attached to a drill, helps remove CO_2 from your mead by agitating it.

Too much dissolved CO_2 can stall fermentation or result in overly fizzy mead. Degassing helps prevent these issues and can also help during clarification stages.

14. Measuring Cups and Scales

Precision is key in mead making. You'll need measuring cups for liquids and a kitchen scale for ingredients like honey, fruit, or additives.

Ensuring you measure ingredients accurately, particularly honey and yeast nutrients, helps maintain consistency between batches and achieve predictable results.

Additional Considerations:

Heat Source

Depending on your recipe, you might need to heat water to dissolve honey or pasteurize ingredients. A kitchen stove or hot plate can serve this purpose.

Cooling Options

Rapidly cooling a hot must can prevent the growth of unwanted bacteria. An ice bath or wort chiller can be handy if you're heating your must.

Labels and Markers

It's useful to label bottles with the date and type of mead to track aging and identify different batches.

Each of these tools and pieces of equipment plays a significant role in the mead-making process, from ensuring the health of your fermentation to maintaining quality control. Beginners can start with the essentials (fermentation vessel, airlock, hydrometer, and bottles) and gradually expand their toolkit as they gain more experience in mead crafting.

Sanitation: Importance and Methods

Mead making, like all fermentation-based craft beverages, hinges on a delicate balance of ingredients, processes, and—perhaps most crucially—sanitation. While honey, water, yeast, and other additives are essential for flavor and structure, none of these elements will result in a successful mead if contamination compromises the fermentation process. We all know how important sanitation is when it comes to fermentation. Sanitation is therefore not just a recommended practice but a foundational step to ensure high-quality mead.

The Importance of Sanitation in Mead Making

Preventing Unwanted Microbial Activity

The process of fermentation involves the controlled activity of yeast, which converts sugars (mainly from honey) into alcohol and carbon dioxide. However, wild yeast, bacteria, and other microorganisms that thrive in the same conditions can spoil a batch of mead by producing off-flavors, mold, or even toxins. Proper sanitation eliminates these unwanted competitors, allowing your chosen yeast strain to flourish without interference.

Preserving Flavor Integrity

Mead is prized for its unique flavor profile, which can range from sweet and floral to dry and spicy, depending on the type of honey, fermentation conditions, and additives used. Even a slight contamination can introduce sour or rancid flavors, masking the delicate characteristics of the mead. By maintaining a clean environment, you preserve the purity of the mead's intended flavor.

Ensuring Consistent Fermentation

Yeast is a sensitive organism. In the presence of contaminants, it may behave unpredictably—fermenting too quickly, stalling entirely, or producing unintended byproducts like fusel alcohols, which can taste harsh or solvent-like. Thorough sanitation ensures that the yeast can perform predictably, resulting in consistent and controlled fermentation from one batch to the next.

Increasing Mead Shelf Life

Contaminants can significantly reduce the shelf life of mead. A batch that hasn't been properly sanitized may develop infections even after bottling, leading to spoilage and potential waste. Effective sanitation minimizes the risk of contamination post-fermentation, helping mead remain stable for months or even years.

Safety Considerations

While it's rare, some contaminants can pose health risks, especially if they involve pathogens or spoilage organisms that produce harmful compounds. Proper sanitation mitigates these risks, ensuring that your mead is not only delicious but also safe to consume.

Methods of Sanitation in Mead Making

Sanitizing in mead making can be divided into two key stages: cleaning and sanitizing. Both are crucial, but they serve different purposes.

Cleaning

Cleaning is the first step, which involves removing visible dirt, residue, or organic material from equipment. Organic matter, such as leftover honey, yeast sediment, or fruit pulp, provides a breeding ground for bacteria and wild yeast. This step doesn't sterilize, but it sets the stage for effective sanitation.

- **Hot Water Rinse**: Begin by rinsing all tools and containers with hot water to loosen any sticky residues or particles. Hot water helps break down honey and other substances that can cling to surfaces.

- **Detergents or Cleaners**: Use an appropriate cleaner to scrub away more stubborn residues. Alkaline-based cleaners like PBW (Powdered Brewery Wash) are commonly used in brewing and mead making because they effectively break down organic material without leaving harmful residues. Avoid using scented or bleach-based household cleaners, as they can leave behind compounds that interfere with fermentation.
- **Mechanical Scrubbing**: For larger surfaces like fermenters or bottles, use brushes or sponges to scrub areas where contaminants might hide, especially in corners, crevices, or around threads in lids and caps.

Sanitizing

After cleaning, sanitizing comes next. Sanitizing doesn't aim to sterilize completely (sterilization refers to killing 100% of microbes), but it reduces microbial life to a safe level so that any remaining organisms cannot compete with the fermentation yeast.

- **Chemical Sanitizers**: The most common method for home mead makers is the use of chemical sanitizers. These are formulated specifically for brewing and food preparation, ensuring they're safe and won't leave harmful residues.
- **Star San**: This is a popular no-rinse sanitizer widely used in brewing. It's acid-based and very effective at killing microbes in as little as 30 seconds. Because it's a no-rinse sanitizer, any foam left behind is safe for contact with the mead and equipment, saving time and reducing the risk of contamination during rinsing.
- **Iodophor**: Another common option, iodophor is an iodine-based sanitizer. It's fast-acting and also safe for brewing, although it may require rinsing depending on the dilution. It is often preferred by mead makers who want an effective and affordable solution.
- **Heat-Based Sanitization**: Some mead makers prefer to sanitize with

heat, particularly for smaller items like bottles or lids. Boiling or steam sterilization is effective at killing bacteria, yeast, and molds. Simply immerse items in boiling water for 15–20 minutes. However, this method isn't suitable for plastic or rubber parts, which can warp or degrade.
- **UV or Ozone Sanitization**: Advanced mead makers or commercial producers might use ultraviolet (UV) light or ozone to sanitize air and surfaces in the brewing environment. These methods are highly effective but are generally used in combination with other techniques to provide an extra layer of protection.

Sanitizing Mead-Making Equipment

- **Fermentation Vessels**: Fermentation vessels are where your mead spends the majority of its time, so it's critical they're thoroughly sanitized. Scrub any residues, paying special attention to seals, taps, and threads. After cleaning, apply a chemical sanitizer and allow it to air dry before use.
- **Airlocks and Bungs**: These are small but critical components, as they provide an entry point for contaminants if not properly sanitized. A quick soak in sanitizer before use will minimize the risk of microbial invasion.
- **Bottles and Caps**: Bottling is a crucial stage where contamination can easily ruin a finished batch. Sanitize bottles by soaking them in sanitizer or running them through a high-temperature dishwasher. Bottle caps or corks should be soaked in sanitizer for a few minutes before use to ensure no contamination during the capping process.

Maintaining a Sanitary Workspace

Your brewing environment is just as important as the equipment itself. If possible, dedicate a clean, clutter-free area for mead making. Surfaces should be wiped down with sanitizing solutions, and any porous materials (like wood or cloth) should be kept away from areas where they might introduce contaminants. Regularly vacuum or clean floors to minimize dust and airborne particulates that could settle in your brew.

Personal Hygiene in Mead Making

As the mead maker, you're also a potential source of contamination. Always wash your hands thoroughly before handling equipment or ingredients, and consider wearing gloves if you're dealing with sensitive stages of the process (like bottling). Wearing a face mask while bottling can help prevent introducing airborne contaminants, particularly if you're working in a less-than-sterile environment.

Setting Up Your Mead Making Space

5

CHAPTER FIVE

PREPARING YOUR MUST

When I first started brewing mead, there was something almost mystical about preparing the must—it's the very lifeblood of your mead, the essence that will transform into a delicious drink over time. To the uninitiated, the term "must" sounds almost too simple for something so important, but it's the foundation on which your mead is built. It's a combination of honey, water, and yeast, blended in just the right way to encourage a magical fermentation process.

Preparing your must is like planting the seeds for a garden; you can't expect a bountiful harvest without careful attention to the soil. Here, your "soil" is a sweet, golden mixture that will soon play host to millions of hungry yeast cells, all working together to create liquid gold.

Step 1: Choosing Your Ingredients

Before you can even think about mixing things up, you need to select the right ingredients. The quality of your honey, water, and yeast will directly impact the final product. Think of it like cooking: you wouldn't use subpar ingredients and expect a gourmet meal. The same holds true for mead.

Honey

I always start with the honey—after all, this is the star of the show. Whether you're going for something light and floral or dark and robust, the honey you choose will define the flavor profile of your mead. In my early days, I used simple clover honey, but over time, I became more adventurous, experimenting with orange blossom honey for a citrusy zing or wildflower honey for a more complex, earthy flavor. You might be tempted to use the cheapest honey you can find, but trust me, it's worth splurging on a quality, raw honey. Pasteurized or overly processed honey will lack the nuances that make your mead unique.

Water

Next comes water, the most overlooked ingredient in mead-making. You'd think water is just water, right? Wrong. Water has a personality, and depending on its mineral content, pH level, and purity, it can subtly influence the outcome of your mead. I usually opt for spring water—something clean, with a neutral pH. If your tap water is chlorinated or heavily treated, it can throw off the flavor or even inhibit fermentation. If in doubt, distilled water is a safe option, but it might lack the mineral complexity that yeast enjoys. And let's not forget, about 70-80% of your must is water, so take it seriously.

Yeast

Then there's yeast, the unsung hero of the fermentation process. It's like the engine driving your mead's transformation from sweet must to alcoholic nectar. For beginners, yeast might seem like a one-size-fits-all deal, but as you delve deeper into mead-making, you'll start to understand the nuances. I've used wine yeasts, champagne yeasts, and even ale yeasts to achieve different results. My go-to for a standard mead is Lalvin D47—it's robust, creates a nice balance of alcohol and sweetness, and leaves a bit of residual honey flavor. Some yeasts can push your mead to dry, others bring out fruity esters, while some allow for higher alcohol tolerance. Yeast selection really depends on what kind of mead you want to craft.

Step 2: Mixing the Must

Once you've gathered your ingredients, it's time to create the must itself. This step feels a bit like alchemy: you're blending together raw components, knowing that, in time, they will transform into something completely new.

First, I start by warming the water—not boiling, but gently heating it to about 100°F (37°C). This step helps dissolve the honey more easily and also ensures a nice environment for the yeast when it's introduced later. Be careful not to overheat the water, as high temperatures can destroy the subtle flavors of your honey and even kill your yeast.

Next, I pour the honey into the warmed water. Depending on your recipe, you'll likely use around 2 to 3 pounds of honey per gallon of water, but this can vary. As I stir, I can't help but feel a connection to ancient mead-makers, who must have done the same thing centuries ago. The honey slowly dissolves, turning the water into a rich, golden syrup. This is your must, the heartbeat of your mead.

At this point, some mead-makers will take a hydrometer reading to measure the specific gravity of the must. This tells you how much sugar is in the mixture and helps estimate the final alcohol content. I like to think of this step as checking the soil before planting—it helps ensure that the yeast will have enough "food" to work with during fermentation.

Step 3: Oxygenation and Nutrients

One of the most important lessons I learned early on is that yeast needs oxygen at the beginning of fermentation. Oxygen helps yeast reproduce and get a strong start before the anaerobic (oxygen-free) environment takes over. In my first few batches, I didn't realize this and ended up with sluggish fermentations. Now, I always vigorously stir or shake the must for a few minutes to introduce oxygen. You'll notice the liquid foaming slightly—this is good!

I also add yeast nutrients at this stage. Honey, despite being packed with sugar, is surprisingly nutrient-poor. Yeast needs more than sugar to thrive—it craves nitrogen and other micronutrients. Depending on the complexity of the must and the yeast strain I'm using, I might add diammonium phosphate (DAP) or a proprietary yeast nutrient mix. Some purists prefer to skip this step, but I've found that it helps ensure a healthy fermentation and prevents off-flavors from stressed yeast.

Step 4: Pitching the Yeast

Now, it's time to introduce the star of the show—yeast. If you're using dry yeast, it's often recommended to rehydrate it first in a small amount of warm water (following the instructions on the packet). This gives the yeast a head start, waking it up from its dormancy. Once rehydrated, I gently pour it into the must, watching as the tiny organisms get to work. If I'm feeling particularly ceremonious, I might give a little nod to the brewing gods at this point—it's tradition, after all!

Step 5: Fermentation

Once the yeast is in, I give the must one final stir and seal it up in my fermentation vessel. Now comes the hard part—waiting. I always imagine the yeast cells bustling about, feasting on sugars, reproducing, and slowly transforming my must into mead. In a few hours, you might notice bubbles forming in the airlock or a frothy layer on top of the must. This is your yeast at work, converting sugars into alcohol and CO_2. The smell changes too, as sweet honey gives way to the tangy scent of fermentation.

Preparing your must is more than just mixing ingredients together—it's the start of a journey, one that requires patience, care, and a little bit of faith in the natural processes at play. Over the years, I've learned to appreciate the subtle art of must preparation, understanding how small changes in water quality, honey type, or even oxygenation can have profound effects on the final mead. Every time I prepare a must, I feel a connection to the countless mead-makers who have come before me, all seeking to turn simple ingredients into something greater than the sum of their parts.

Measuring and Mixing Ingredients

The balance of honey, water, yeast, and nutrients determines the final product—whether you end up with a sweet, golden ambrosia or something that tastes more like vinegar. I've come to realize that while mead-making allows for creativity, there is a science behind it, and precision in measuring and mixing your ingredients is essential to ensure a successful batch.

In my early days of brewing, I took a relaxed, almost casual approach to measuring. I figured mead was an ancient drink, made by Vikings and medieval monks, so how exact could it really be? As it turns out, the answer is *very*. While mead-making may seem simple—just honey, water, and yeast—the ratio of these ingredients plays a critical role in how your mead will ferment and ultimately taste.

The three core elements—honey, water, and yeast—each have specific roles to play, and if you misjudge the proportions, it can lead to a host of problems. Too little honey, and your mead will be weak and dry. Too much honey, and fermentation could stall. Incorrect amounts of yeast or improper nutrient additions could result in off-flavors or incomplete fermentation.

Measuring Honey

Honey is the heart and soul of your mead, and getting the right amount is crucial. The amount of honey you use directly impacts the sweetness, alcohol content, and body of your mead. Typically, mead-makers use anywhere from 2 to 5 pounds of honey per gallon of water, depending on the style of mead they want to create. For a standard mead (or *traditional* mead, as it's often called), I usually aim for about 3 pounds of honey per gallon.

To measure honey accurately, I use a kitchen scale. Honey is heavy and viscous, and measuring by volume can be tricky and inconsistent. Weighing it ensures precision. I learned quickly that pouring honey straight from the jar can be a sticky, messy business. To make things easier, I often warm the jar in a pot of hot water for a few minutes to loosen the honey and make it easier to pour. A warm jar of honey flows like liquid gold, filling your kitchen with that sweet, floral scent that promises something delicious is in the works.

Measuring Water

Next comes the water. Water seems simple enough, but it plays a vital role in balancing the must and affecting the fermentation process. While honey provides the sugars that the yeast will feast on, water makes up the majority of the liquid volume and is the medium in which all the magic happens.

I've found that the best approach is to measure water by volume, typically using a standard gallon jug or measuring cup. The standard ratio for a traditional mead is roughly 1 gallon of water per 3 pounds of honey. But, like any recipe, these ratios can shift based on the mead you're aiming for. For example, if you're making a hydromel (a lighter, more sessionable mead), you might use less honey, say around 2 pounds per gallon, to keep the alcohol content and sweetness lower.

Using clean, filtered water is essential. I've learned to avoid tap water, as it often contains chlorine or other chemicals that can affect fermentation and flavor. Instead, I opt for spring water, which has a more natural mineral balance and won't interfere with the yeast. I've also played around with distilled water, which works but can leave your must feeling a bit flat if it lacks essential minerals that yeast enjoy.

Measuring Yeast

Yeast, despite its tiny size, has a massive role in how your mead turns out. It's the yeast that consumes the sugars in honey, converting them into alcohol and CO_2. Getting the right amount of yeast is crucial—not enough, and the fermentation will be sluggish or incomplete; too much, and the yeast may produce too many off-flavors, especially if they get stressed.

Most mead recipes call for about 1 gram of yeast per gallon of must. This may seem like a small amount, but yeast is a powerful organism. When I first started, I would eyeball the amount of yeast, sprinkling it into the must like seasoning on a meal. I quickly learned that this was a bad idea—too little yeast, and I ended up with mead that took weeks to show any sign of fermentation. Now, I measure it precisely with a small kitchen scale or use a packet of pre-measured yeast, which simplifies things.

Measuring Nutrients

One thing that surprises many first-time mead-makers is that honey, while packed with sugar, is actually nutrient-poor. Yeast needs more than sugar to thrive; it also requires nitrogen, vitamins, and minerals. Without these nutrients, the yeast can become stressed, leading to off-flavors like rotten eggs or sulfur, a mistake I encountered in one of my early batches.

There are different ways to add nutrients, and the exact amount depends on the specific strain of yeast you're using and the type of mead you're making. A common rule of thumb is to add around 1 teaspoon of yeast nutrient per gallon of must, but it's always a good idea to follow the guidance provided by your yeast manufacturer. You can also split the nutrient additions into staggered intervals during fermentation, which helps the yeast stay healthy throughout the process.

CHAPTER FIVE

Once your ingredients are measured, it's time to mix them together to form your must. Mixing is deceptively simple, but it's another step where attention to detail matters. The goal is to fully dissolve the honey into the water and introduce oxygen into the must before fermentation begins.

I like to start by gently warming the water to about 100°F (37°C). This makes it easier to incorporate the honey without damaging its delicate flavors. You don't want to heat the water too much—never to boiling—as that can degrade the honey and kill off beneficial enzymes and wild yeasts present in raw honey.

Like I said earlier, I pour the honey into the warm water and stir vigorously, ensuring that it dissolves completely. This can take a little time, depending on how thick your honey is. As I stir, I make sure to introduce plenty of oxygen into the must, which is essential for the yeast during the first phase of fermentation. Oxygen helps the yeast reproduce and establish a strong colony before they shift to anaerobic fermentation, where they'll start producing alcohol.

Once the honey is fully mixed in and the must is smooth and uniform, I go with the yeast and add the nutrients. Then, I give the whole mixture a good stir before transferring it into the fermentation vessel.

Understanding the Role of pH and Acidity

When I first started making mead, the word "pH" felt like it belonged more in a chemistry lab than in my kitchen. I knew it had something to do with acidity, but its significance in mead-making wasn't immediately obvious. Over time, I learned that pH and acidity play a crucial role in how your mead ferments, tastes, and matures. If you want to take your mead-making to the next level, understanding and managing pH is just as important as selecting the right honey or yeast.

What is pH and Why Does It Matter?

At its core, pH measures how acidic or basic a solution is on a scale from 0 to 14, with 7 being neutral. Solutions with a pH lower than 7 are acidic (like lemon juice), while those with a pH higher than 7 are basic or alkaline (like baking soda). Mead, like most fermented beverages, falls on the acidic side of the scale, typically somewhere between 3 and 4.5 during fermentation.

So, why does pH matter in mead-making? The answer lies in its influence over several critical aspects of the fermentation process, including yeast health, fermentation speed, and flavor development. pH levels can make or break your mead, and striking the right balance can lead to a smooth, well-rounded drink. Ignore it, and you might end up with a stalled fermentation, unpleasant flavors, or mead that ages poorly.

The Role of Acidity in Fermentation

Yeast Health and Fermentation Activity

Yeast, the little organisms responsible for transforming sugars into alcohol, thrive within a specific pH range. For most strains used in mead-making, the optimal pH falls between 3.7 and 4.5. This range provides an ideal environment for yeast to function without becoming stressed. If the pH drops too low—below 3.0—the yeast can become sluggish or even stop fermenting altogether, a frustrating experience I encountered early on when I neglected pH management.

I vividly remember one of my first batches where fermentation simply stalled. I'd done everything else right—measured my honey, added yeast nutrients, and kept the fermentation temperature steady. But when I tested the pH, I was shocked to see it had dropped below 3.0. The yeast was essentially in a hostile environment, and fermentation had slowed to a crawl.

A small pH adjustment brought it back to life, but I learned a valuable lesson: pH can make or break your fermentation.

Buffering Against Stalled Fermentation

Mead-making often starts with a must that is naturally low in nutrients compared to beer or wine. Honey is an amazing source of sugar, but it's not exactly brimming with the vitamins and minerals yeast need to stay healthy. This can lead to stressed yeast, which may produce unwanted off-flavors like sulfur or, in the worst cases, halt fermentation altogether. Stressed yeast are also more likely to struggle in environments with a lower pH.

That's where buffering comes in. A good buffering system helps stabilize the pH levels during fermentation, preventing drastic drops that can harm the yeast. One common method is the addition of potassium bicarbonate or calcium carbonate, which can raise the pH if it becomes too acidic. However, adding these compounds requires caution—too much can swing the pH too far in the other direction, leading to a basic environment where yeast won't thrive either.

Acidity and Flavor Balance

Acidity doesn't just affect yeast health; it also has a profound influence on the flavor of your mead. Acidity adds brightness and structure to the drink, balancing the sweetness of the honey and making the final product more complex and enjoyable to drink. Without the right level of acidity, your mead can taste flat, cloyingly sweet, or lifeless.

A mead with well-balanced acidity will have a crisp, refreshing taste that complements the honey's natural sweetness. When acidity is too low, your mead may lack that refreshing quality, leaving behind a syrupy, one-dimensional flavor. On the other hand, if acidity is too high, your mead could taste harsh or overly tart, masking the subtle floral and fruity notes you want to highlight.

I've found that acidity can make the difference between a good mead and a great one. One of my favorite batches used a blend of orange blossom honey, which has a naturally lighter, more citrusy profile. By carefully monitoring the pH and maintaining the right acidity, the final mead had a delicate balance between sweet and tart, with a complexity that made it stand out from my other batches.

Monitoring and Adjusting pH in Mead-Making

Now that we understand the importance of pH, how do you monitor and adjust it? Thankfully, this isn't as complicated as it might seem. Here are the steps I follow to ensure the pH of my mead is in the optimal range for fermentation and flavor development:

1. Testing pH Levels

To keep an eye on pH levels, I use a pH meter. While pH strips are an option, I've found them to be less accurate, especially when trying to measure the subtle differences within the 3.0 to 4.5 range. A decent pH meter is a small investment that will pay off in consistently good batches of mead.

I test the pH of my must before pitching the yeast to ensure it's in the ideal range (usually around 3.8 to 4.2), and then I monitor it periodically throughout the fermentation process. If I notice the pH starting to drop below 3.0, that's a red flag. At this point, I'll take action to adjust it.

2. Adjusting pH

If the pH is too low, I'll use a small amount of potassium bicarbonate or calcium carbonate to raise it. I prefer potassium bicarbonate because it's more gentle and less likely to overshoot the desired pH range. I've learned through trial and error that adding too much at once can send the pH spiraling in the opposite direction, so I always start small—adding just a pinch, stirring, and retesting before adding more.

It's important to be patient with pH adjustments. You don't want to make large changes all at once, as this can stress the yeast and disrupt the fermentation process. I prefer to check the pH regularly and make gradual corrections over the course of a day or two if necessary.

3. Acidity in the Finished Product

While pH is critical during fermentation, the acidity in the finished product also plays a role in how your mead is perceived. After fermentation is complete, I taste my mead to assess whether it needs any adjustments. If it's too sweet or flat, I might add a touch of acid, such as tartaric acid or citric acid, to balance the flavor. I do this sparingly—just enough to brighten the mead without overwhelming the delicate flavors of the honey.

Adjusting Sugar Levels: Hydrometer Use and Potential Alcohol Content

When it comes to mead-making, sugar is not just a sweetener; it's the very fuel that drives fermentation, feeding the yeast and ultimately determining the alcohol content. At the heart of controlling sugar levels—and thereby influencing the final taste and strength of your mead—is the trusty hydrometer. This little tool quickly became one of the most valuable pieces of equipment in my mead-making journey. Understanding how to use a hydrometer and adjust sugar levels is critical for crafting the perfect batch, especially when you're aiming for specific alcohol content or sweetness levels.

The Role of Sugar in Mead-Making

Before we talk about how technical use of hydrometer is, let's start with why sugar is so essential in mead-making. Honey, the main sugar source in mead, provides fermentable sugars that yeast converts into alcohol. The amount of sugar present in your must (the mixture of honey, water, and other ingredients) directly impacts three things: the alcohol content, sweetness, and body of the final product.

Too little sugar, and your mead will be weak, both in flavor and alcohol content. Too much sugar, and the yeast might struggle to ferment all of it, leaving behind unfermented sugars that make your mead overly sweet—or worse, causing fermentation to stall. It's all about balance, and that's where the hydrometer comes into play.

What is a Hydrometer and How Does It Work?

As a beginner in brewing, the idea of using a hydrometer seemed intimidating. It looked like a fragile glass thermometer floating in liquid, and I wasn't quite sure what to do with it. But once I understood how it worked, it became my go-to tool for tracking fermentation progress and adjusting sugar levels.

A hydrometer measures the density, or *specific gravity* (SG), of a liquid. In the context of mead-making, this means it measures how much sugar is dissolved in your must compared to water. Pure water has a specific gravity of 1.000. As you dissolve honey in water, the SG rises because honey is denser than water. After fermentation, when the yeast has consumed most of the sugars, the SG will drop, sometimes even below 1.000, depending on how dry your mead is.

The beauty of the hydrometer is that it gives you a snapshot of where your must is at any given time during the fermentation process. It helps you determine whether there is enough sugar for fermentation, whether

CHAPTER FIVE

fermentation is proceeding as expected, and what the potential alcohol content will be once fermentation is complete.

Step 1: Taking Initial Readings

One of the first things I do after mixing my honey and water into must is take an initial hydrometer reading, which tells me the starting specific gravity (SG). This initial reading is critical because it allows you to estimate the potential alcohol content of your mead. If your starting gravity is too low, your mead will have a low alcohol content (and potentially taste thin). If it's too high, you might end up with an overly sweet mead, or the yeast might not be able to fully ferment all the sugar.

I start by filling my testing tube with the must, making sure there are no bubbles or chunks of honey floating around. Then, I gently place the hydrometer into the tube, giving it a spin to dislodge any bubbles that might affect the reading. When the hydrometer settles, I take note of where the liquid meets the scale on the hydrometer—this is my initial specific gravity.

For most meads, I aim for an initial specific gravity between 1.080 and 1.120. A lower number will give you a lighter, lower-alcohol mead, while a higher number can produce a stronger, sweeter beverage. The general rule is that each 0.001 SG corresponds to roughly 0.13% alcohol by volume (ABV), so a starting gravity of 1.100, for example, would have the potential to produce around 13% ABV.

Step 2: Calculating Potential Alcohol Content

The potential alcohol content of your mead is directly related to how much sugar is in the must at the start of fermentation. Luckily, with a hydrometer, calculating potential alcohol is relatively straightforward.

Here's how I do it: I subtract the final specific gravity (when fermentation has stopped) from the starting specific gravity, and then multiply the difference by 131. This gives me the approximate alcohol by volume (ABV) of the mead. The formula looks like this:

$$ABV = (\text{Starting Gravity} - \text{Final Gravity}) \times 131$$

For example, if I start with a specific gravity of 1.100 and the fermentation finishes at 1.000, my ABV calculation would be:

$$(1.100 - 1.000) \times 131 = 13.1\% \text{ ABV}$$

This simple formula allows me to gauge the strength of the mead and adjust my recipe accordingly. If I'm aiming for a lighter, sessionable mead, I might lower the starting gravity by using less honey. If I want a more potent drink, I'll increase the honey content.

Step 3: Adjusting Sugar Levels

There are times when the initial reading isn't quite where I want it to be, either because I've misjudged the honey-to-water ratio or I've decided on a different target alcohol content. The good news is that adjusting sugar levels at this stage is easy.

If my starting gravity is too low, I'll add more honey to increase the sugar content and bring the gravity up. I do this in small increments, stirring thoroughly and testing again with the hydrometer after each addition. It's better to add honey gradually than to overshoot and end up with a must that's too sweet.

If the starting gravity is too high, I'll dilute the must with a little more water to bring it down. Again, I do this in small amounts, retesting after each addition. It's important to remember that once fermentation starts, adjusting the sugar levels becomes much more difficult, so it's best to get things right at the start.

Step 4: Monitoring Fermentation Progress

Once fermentation begins, the hydrometer becomes a valuable tool for tracking its progress. Yeast consumes sugar and produces alcohol, which lowers the specific gravity over time. By taking regular hydrometer readings—usually every week—I can see how much sugar remains and estimate how much longer fermentation will take.

A steady drop in SG readings is a good sign that fermentation is proceeding as it should. If the specific gravity stops decreasing and stabilizes at a number above 1.000, I know that fermentation is complete or close to it. I also look out for signs of a stuck fermentation, which can happen if the yeast gets stressed or runs out of nutrients. If the SG remains the same over several readings but fermentation hasn't finished, I might need to step in and adjust the environment to help the yeast along.

Step 5: Adjusting Sweetness After Fermentation

Once fermentation has finished, I take my final hydrometer reading and calculate the alcohol content. At this point, I also assess the sweetness of the mead. Sometimes, even after fermentation has gone smoothly, the mead might turn out drier than I expected. If the final specific gravity is below 1.000, the mead will likely be quite dry, with little to no residual sugar.

To adjust the sweetness, I can back-sweeten the mead by adding additional honey or another sugar source. However, it's essential to stabilize the mead with potassium sorbate and potassium metabisulfite before back-sweetening, as this prevents the yeast from reactivating and fermenting the added sugars. Once stabilized, I add honey in small increments, stirring and tasting until I reach the desired sweetness.

Adding Nutrients: Importance of Yeast Nutrients and Feeding Schedules

Why Yeast Nutrients are Important

Let's start with the basics: yeast are living organisms. Like any living thing, they need more than just sugar to survive and do their job. In mead-making, we rely on yeast to convert the sugars in honey into alcohol and carbon dioxide. But honey, as wonderful as it is, doesn't naturally contain the vitamins, minerals, nitrogen, or other essential nutrients yeast need to stay healthy and active during fermentation.

Without these nutrients, the yeast can become stressed. And stressed yeast are bad news. They can produce off-flavors like sulfur, or they might simply stop fermenting altogether, leaving you with an overly sweet mead that's stuck halfway through fermentation. Worse yet, they may create unwanted byproducts that take months to age out, if they ever do. So, providing yeast with a balanced diet through proper nutrient additions is essential to ensure a clean, steady fermentation.

Types of Yeast Nutrients

When I started making mead, I was overwhelmed by the different types of yeast nutrients available. There are quite a few options, but they all serve the same goal: keeping your yeast healthy. Here's a breakdown of the most commonly used nutrients in mead-making:

Diammonium Phosphate (DAP)

DAP is a source of inorganic nitrogen, which is crucial for yeast metabolism. Nitrogen helps yeast build proteins and enzymes necessary for fermentation. However, too much DAP can lead to off-flavors, so it's important to use it in moderation and in combination with other nutrients.

Fermaid O

Fermaid O is an organic yeast nutrient made from inactivated yeast cells. It provides a source of organic nitrogen, as well as essential vitamins, amino acids, and fatty acids. The organic nitrogen in Fermaid O is more slowly absorbed by yeast, which makes it a great option for a controlled, steady release of nutrients.

Fermaid K

This contains a mixture of nitrogen (both inorganic and organic), as well as important nutrients like magnesium, B vitamins, and minerals. It's an all-in-one nutrient blend that can support yeast health throughout fermentation.

Yeast Hulls

Yeast hulls, or dead yeast cells, are often used as a last-resort nutrient addition when fermentation stalls. They provide nutrients, vitamins, and sterols that can help revive sluggish yeast by absorbing toxins and providing a buffer.

Go-Ferm

Go-Ferm is typically used during yeast rehydration. It contains vitamins, minerals, and micronutrients that help yeast acclimate to the fermentation environment and get off to a strong start. It's especially useful when working with high-sugar musts or aggressive fermentation conditions.

Yeast Nutrients and the Nitrogen Factor

Nitrogen is a key component in yeast nutrition, but not all nitrogen is the same. Yeast require both organic nitrogen (found in things like Fermaid O) and inorganic nitrogen (found in DAP). The balance of these two types of nitrogen is crucial. Organic nitrogen is slowly assimilated, providing a steady source of nutrition throughout fermentation, while inorganic nitrogen, like DAP, is rapidly consumed early on.

Inadequate nitrogen can cause the yeast to become stressed, leading to the production of unwanted compounds such as hydrogen sulfide (the dreaded "rotten egg" smell). On the other hand, too much nitrogen can also cause problems, including unwanted flavors. This is why a balanced nutrient schedule that includes both types of nitrogen is essential.

The Importance of a Feeding Schedule

Nutrient timing is everything. When I first started making mead, I made the mistake of dumping all the nutrients into the must right at the beginning. While this seemed like a simple, efficient approach, it can actually cause more harm than good. Adding too many nutrients at once can lead to an aggressive fermentation that quickly dies out, leaving behind yeast that struggle to finish the job.

Yeast need nutrients throughout fermentation, not just in the initial stages. This is where a feeding schedule comes in. By dividing the nutrient additions into smaller doses over time, you can ensure that the yeast have a steady supply of what they need, when they need it. This helps maintain a consistent fermentation and reduces the risk of off-flavors or stalled fermentation.

The most common method for managing nutrient additions is called *staggered nutrient additions* (SNA). Here's how it works:

Initial Addition (Pitching the Yeast)

When you pitch your yeast into the must, the first nutrient addition is typically made. This provides the yeast with the initial boost they need to kick-start fermentation. I usually add a combination of Go-Ferm (during yeast rehydration) and Fermaid O for organic nitrogen.

First 24 Hours

About 24 hours after pitching the yeast, I make the second nutrient addition. By this point, the yeast have consumed some of the initial nutrients, and adding more keeps them fed as they continue to multiply and ferment the sugars in the must. At this stage, I'll often add a combination of DAP and Fermaid O to balance the organic and inorganic nitrogen sources.

48-72 Hours

After another day or two, I check the gravity and the activity of fermentation. This is usually when fermentation is at its most vigorous, and the yeast are rapidly converting sugars into alcohol. I'll add another small dose of nutrients—usually a mix of Fermaid K and DAP—to ensure the yeast have enough nitrogen to complete the job.

Halfway Through Fermentation

Once the specific gravity has dropped by about half of the original reading, I add the final nutrient dose. This usually happens around 4-5 days into fermentation. By this time, the yeast are settling into the later stages of fermentation, and this final nutrient addition helps them finish strong without becoming stressed.

Avoiding Overfeeding

One of the challenges of using yeast nutrients is knowing when to stop. While yeast need nutrients to thrive, overfeeding them can result in off-flavors, over-attenuation, or excessive foaming. Mead that's too rich in nutrients may ferment too quickly, leading to a loss of delicate honey flavors or unintentional dryness. Additionally, too much DAP can lead to harsh chemical flavors, so it's important to stick to a measured approach and adjust as necessary.

I've learned to watch the yeast's activity closely, making adjustments as needed. If fermentation seems overly aggressive, I might dial back the nutrient additions. On the other hand, if fermentation slows unexpectedly, I might add a small dose of yeast hulls to help absorb any toxins and give the yeast a little extra support.

The Long-Term Impact of Proper Nutrition

The benefits of proper nutrient management go beyond just getting through fermentation. By feeding yeast correctly, you set the stage for a cleaner, more stable mead. Healthy yeast will ferment more efficiently, leaving behind fewer off-flavors that need to age out over time. The final product will also be more predictable, with a more balanced alcohol and sweetness profile.

In some of my earlier batches, I was impatient, adding all the nutrients at once and hoping for the best. The result was often an overly sweet, under-fermented mead that took months (if not years) to mellow out. Once I started using staggered nutrient additions, I noticed a huge improvement. Fermentation was smoother, and the flavors were cleaner right from the start, meaning I could enjoy my mead much sooner without waiting for unpleasant flavors to fade away.

6

CHAPTER SIX

FERMENTATION

Fermentation in mead making is the process where yeast converts sugars from honey into alcohol and carbon dioxide. After mixing honey with water, yeast is introduced to start the fermentation. This process typically takes weeks to months, depending on factors like temperature and yeast strain. As yeast consumes the sugars, it produces alcohol, resulting in the transformation of the honey-water mixture into mead. Fermentation can be influenced by nutrients, pH levels, and oxygen management, all of which affect the flavor, alcohol content, and clarity of the final product.

Primary vs. Secondary Fermentation: Processes and Differences

The fermentation process typically occurs in two stages: primary and secondary. Each stage plays a crucial role in developing the flavor, clarity, and overall character of the mead. Understanding the differences between these two stages is essential for producing a high-quality beverage, as each phase has distinct purposes and characteristics.We will look into the processes and differences between primary and secondary fermentation in mead making, focusing on how each step impacts the final product.

The Primary Fermentation Process

Primary fermentation is the initial stage where the bulk of the yeast activity takes place. After mixing honey with water, known as "must," and introducing yeast, the fermentation process begins. During this phase, yeast cells rapidly multiply and start converting the sugars in honey into alcohol and carbon dioxide. The environment during primary fermentation is oxygen-rich, which helps the yeast population to grow before they switch to anaerobic fermentation (alcohol production).

This stage typically lasts anywhere from one to three weeks, depending on factors such as temperature, yeast strain, and the sugar content of the must. At the peak of primary fermentation, the must will be actively bubbling as carbon dioxide is produced. This is a vigorous phase, where most of the alcohol is generated, and the yeast consumes a significant portion of the sugars.

One of the main purposes of primary fermentation is to establish the alcohol content of the mead. However, it also begins the initial development of flavor. During this stage, the yeast's metabolism and byproducts can impart certain flavors to the mead, such as fruity esters or subtle phenolics, which contribute to the complexity of the beverage.

The Secondary Fermentation Process

Once primary fermentation has slowed down and the bubbling subsides, the mead enters secondary fermentation. This is the point at which most of the sugars have been converted, and the yeast activity begins to taper off. Secondary fermentation is often referred to as the maturation phase, where mead is left to age, clarify, and develop more nuanced flavors.

The transfer from primary to secondary fermentation usually involves "racking," a process where the mead is siphoned off from the sediment, or lees, that has settled at the bottom of the fermentation vessel. This lees consists of dead yeast cells, proteins, and other particulate matter that can contribute off-flavors if left too long in contact with the mead. Racking helps clarify the mead and improves its taste and appearance.

During secondary fermentation, oxygen exposure is minimized to prevent oxidation, which can spoil the mead. Yeast activity during this phase is much slower, and little to no carbon dioxide is produced. Secondary fermentation can last anywhere from a few weeks to several months, depending on the mead maker's goals. This phase allows for the development of more complex flavors, as the mead continues to age and mellow. Any harsh alcohol flavors produced during primary fermentation begin to soften, and the mead takes on a smoother, more refined character.

Key Differences Between Primary and Secondary Fermentation

Yeast Activity

The most significant difference between primary and secondary fermentation is the level of yeast activity. In primary fermentation, yeast is highly active, consuming sugars and producing alcohol and carbon dioxide at a rapid pace. Secondary fermentation, on the other hand, sees a dramatic slowdown in yeast activity as most of the sugars have already been converted.

Duration

Primary fermentation is relatively short, typically lasting between one and three weeks, while secondary fermentation can take several weeks to months. The length of secondary fermentation depends on the mead style and the desired final product, as aging during this phase significantly impacts the mead's complexity and smoothness.

Purpose

The primary goal of primary fermentation is alcohol production, while secondary fermentation focuses on maturation, flavor development, and clarification. The two processes complement each other: primary fermentation creates the base alcohol and some initial flavors, while secondary fermentation refines those flavors and improves the mead's overall quality.

Environment

Primary fermentation occurs in an oxygen-rich environment to encourage yeast growth, while secondary fermentation requires an oxygen-free environment to prevent spoilage. Oxygen exposure during secondary fermentation can lead to oxidation, which causes undesirable flavors like staleness or a "wet cardboard" taste.

Sediment Management

During primary fermentation, yeast and other solids (such as proteins) fall out of suspension, creating sediment at the bottom of the fermentation vessel. While this is a natural part of the process, it is essential to rack the mead into a clean vessel during secondary fermentation to avoid off-flavors caused by prolonged contact with the lees.

Clarification and Stabilization

Mead typically looks cloudy during primary fermentation due to yeast and particulate matter in suspension. In secondary fermentation, the mead gradually clears as yeast activity slows and particles settle out. Clarifying agents can be added during this phase to further improve clarity. Additionally, secondary fermentation is the stage where the mead stabilizes, as any remaining sugars are fully converted, reducing the risk of unintentional re-fermentation when the mead is bottled.

The Importance of Both Stages in Mead Making

Both primary and secondary fermentation are critical to the mead-making process. Skipping or rushing through either phase can result in an imbalanced, cloudy, or off-flavored final product. Primary fermentation is essential for alcohol production and sets the foundation for the mead's flavor. Secondary fermentation, while slower and less active, is equally important for refining the mead, developing subtle flavors, and achieving clarity and smoothness.

The timing of when to move from primary to secondary fermentation is crucial. If mead is transferred too early, yeast may not have finished converting all the sugars, leading to a stuck fermentation later. If left too long in primary fermentation, however, the mead can develop off-flavors from the lees, impacting its taste and quality.

Managing Fermentation: Temperature, Duration, and Signs of Progress

Effective management of fermentation conditions is essential to ensure a successful outcome. Key aspects of fermentation management include controlling temperature, understanding duration, and recognizing signs of progress. This essay explores these elements in detail, highlighting their impact on the fermentation process and the quality of the final mead.

Temperature Control

Temperature is one of the most critical factors in managing fermentation. It affects yeast activity, flavor development, and overall fermentation efficiency.

Yeast Activity

Yeast strains have specific temperature ranges in which they operate optimally. For most mead yeasts, this range is between 60°F and 75°F (15°C to 24°C). Within this range, yeast cells metabolize sugars efficiently, producing alcohol and carbon dioxide at a steady rate. If the temperature is too low, yeast activity slows down, potentially leading to a stuck fermentation. Conversely, if the temperature is too high, yeast may become overly active, producing excessive heat and byproducts like fusel alcohols and esters, which can result in off-flavors.

Flavor Development

The temperature also influences the flavor profile of the mead. Cooler temperatures tend to produce cleaner, more subdued flavors, while warmer temperatures can enhance ester and phenol production, leading to more pronounced fruity or spicy notes. Therefore, maintaining a consistent temperature within the recommended range is vital for achieving the desired flavor profile.

Fermentation Vessel

The type of fermentation vessel and its insulation properties can also impact temperature control. For instance, glass carboys and plastic fermenters can have different heat retention properties. Using a temperature-controlled environment, such as a fermentation chamber or temperature-controlled room, helps maintain the optimal conditions for yeast activity.

Duration of Fermentation

The duration of fermentation can significantly impact the final quality of the mead.

Primary Fermentation

Primary fermentation typically lasts between one and three weeks. The duration depends on factors such as the yeast strain, initial sugar concentration, and fermentation temperature. It is essential to allow sufficient time for the yeast to complete its work. Rushing the process can result in residual sugars, incomplete alcohol production, and off-flavors.

Secondary Fermentation

Once primary fermentation is complete, mead transitions to secondary fermentation, which can last several weeks to several months. This stage is crucial for aging and developing the mead's flavor profile. The extended duration allows for the maturation of flavors and the clarification of the mead. Patience during secondary fermentation is key to achieving a smoother and more refined final product.

Monitoring Progress

Regularly monitoring the specific gravity of the mead using a hydrometer or refractometer helps determine the fermentation progress. A stable specific gravity over several days indicates that fermentation is complete. This practice prevents premature bottling, which can result in bottle bombs if residual sugars continue to ferment.

Signs of Progress

Recognizing the signs of fermentation progress is essential for effective management and ensuring the quality of the mead.

Bubbling and Airlock Activity

During primary fermentation, the most obvious sign of progress is bubbling in the airlock or fermentation vessel. This indicates the production of carbon dioxide as yeast consumes sugars. As fermentation nears completion, bubbling will slow down and eventually stop.

Specific Gravity Readings

Measuring the specific gravity provides a quantitative indication of fermentation progress. Initial specific gravity readings are taken before fermentation begins, and subsequent readings are taken periodically. A decrease in specific gravity signifies that the yeast is converting sugars into alcohol. Once the gravity stabilizes at a low level, fermentation is considered complete.

Sediment Formation

The formation of sediment, or lees, at the bottom of the fermentation vessel is another sign that fermentation is progressing. As yeast and other particles settle out, it indicates that the fermentation process is underway. In primary fermentation, this sediment will accumulate, and in secondary fermentation, it will compact as the mead clarifies.

Taste and Smell

Observing changes in taste and smell can also provide insights into fermentation progress. During primary fermentation, the mead may have a yeasty or fruity aroma, which will diminish as fermentation completes. Tasting the mead periodically helps assess the development of flavors and identify any potential issues.

Troubleshooting Common Fermentation Issues

Despite careful planning and execution, mead makers may encounter various fermentation issues that can impact the final product's quality. Troubleshooting these issues requires an understanding of potential problems and their causes. Here, we shall explore common fermentation issues in mead making, their causes, and strategies for resolution.

1. Stuck Fermentation

A stuck fermentation occurs when yeast activity halts before all sugars are converted into alcohol, leaving the mead with a high specific gravity and unfermented sweetness.

Causes:

- **Temperature Extremes**: Yeast can become inactive in temperatures that are too low or too high.
- **Nutrient Deficiency**: Yeast requires certain nutrients to thrive, including nitrogen, which can be depleted.
- **High Sugar Concentration**: Excessively high sugar levels can create an environment where yeast struggles to function.

Resolution:

- **Adjust Temperature**: Ensure the fermentation temperature is within the optimal range for the yeast strain used, typically between 60°F and 75°F (15°C to 24°C).
- **Add Yeast Nutrients**: If nutrient deficiency is suspected, adding a yeast nutrient blend can help revive fermentation.
- **Rehydrate Yeast**: If fermentation has stalled due to high sugar concentrations, consider rehydrating additional yeast to boost activity.

2. Off-Flavors and Aromas

Off-flavors or undesirable aromas in mead can manifest as solvent-like smells, overly fruity or spicy notes, or a musty or unpleasant taste.

Causes:

- **Temperature Fluctuations**: Rapid changes in temperature can stress yeast, leading to off-flavors.
- **Poor Sanitation**: Contamination from bacteria or wild yeast can introduce unwanted flavors.
- **Oxygen Exposure**: Excessive oxygen during fermentation can cause oxidation, leading to stale or off-flavors.

Resolution:

- **Maintain Stable Temperature**: Keep fermentation temperatures consistent and within the recommended range for the yeast strain.
- **Sanitize Equipment**: Ensure all fermentation equipment is thoroughly cleaned and sanitized to prevent contamination.
- **Minimize Oxygen Exposure**: Use airlocks and avoid splashing to reduce the risk of oxidation.

3. Excessive Sediment

Excessive sediment, or lees, is a thick layer of yeast and other particles that accumulate at the bottom of the fermentation vessel. While some sediment is normal, excessive amounts can lead to off-flavors and clarity issues.

Causes:

- **Overly Aggressive Fermentation**: Rapid fermentation can result in an excessive amount of yeast and byproducts.
- **Inadequate Racking**: Failing to transfer the mead off the sediment can lead to prolonged contact with the lees.

Resolution:

- **Rack the Mead**: Transfer the mead to a clean vessel, leaving the sediment behind. This process, known as racking, helps clarify the mead and prevents off-flavors from developing.
- **Use Fining Agents**: Consider using fining agents to help clarify the mead and remove excess sediment.

4. Low Alcohol Content

Low alcohol content can result if the yeast fails to fully ferment the sugars or if the initial sugar concentration is too low.

Causes:

- **Incomplete Fermentation**: If yeast activity is insufficient, the fermentation may not reach the desired alcohol level.
- **Initial Sugar Level**: If the initial sugar concentration is too low, the resulting alcohol content will also be low.

Resolution:

- **Check Yeast Health**: Ensure that yeast is viable and properly pitched. Re-pitching fresh yeast or adding yeast nutrients can help achieve complete fermentation.
- **Adjust Honey Content**: If the initial sugar concentration was low, consider adjusting the honey content in future batches to reach the desired alcohol level.

5. Cloudiness and Lack of Clarity

Cloudiness in mead can be an aesthetic issue and may indicate incomplete fermentation or the presence of suspended particles.

Causes:

- **Inadequate Settling Time**: Mead may still be cloudy if it has not had enough time to settle and clarify.
- **Presence of Proteins or Pectins**: Certain proteins or pectins from honey can cause cloudiness.

Resolution:

- **Allow Time to Clear**: Give the mead additional time to settle and clarify naturally. This can take several weeks to months.
- **Use Clarifying Agents**: Consider using fining agents, such as bentonite

or gelatin, to help clarify the mead. Follow the recommended usage instructions to avoid overuse.

6. Over-Carbonation

Over-carbonation occurs when excess carbon dioxide is produced, leading to excessive fizz or carbonation in the final mead.

Causes:

- **Incomplete Fermentation**: If fermentation is not complete before bottling, residual sugars can cause additional carbonation.
- **Improper Bottling**: Adding priming sugars or bottling before fermentation is fully finished can lead to over-carbonation.

Resolution:

- **Wait for Complete Fermentation**: Ensure that fermentation is complete before bottling. Monitor specific gravity readings to confirm that fermentation has finished.
- **Control Priming Sugar**: If using priming sugars, calculate the appropriate amount based on the volume of mead and desired carbonation level.

RACKING

Racking is a fundamental process in mead making, crucial for achieving a clear, refined, and well-balanced beverage. This technique involves transferring mead from one vessel to another, leaving sediment behind, and is an essential step for ensuring the final product is free from undesirable particles and flavors. This essay explores the racking process in detail, including its purpose, techniques, best practices, and its impact on the quality of the mead.

Purpose of Racking

Racking serves several important purposes in the mead-making process:

Clarification

One of the primary goals of racking is to improve the clarity of the mead. During fermentation, yeast and other particles, such as proteins and tannins, settle at the bottom of the fermentation vessel. Racking removes the mead from this sediment, reducing cloudiness and enhancing visual appeal.

Flavor Development

Extended contact with sediment, or lees, can introduce off-flavors into the mead, such as yeasty or sulfury tastes. By racking the mead, the risk of these undesirable flavors is minimized, allowing for a cleaner and more refined taste profile.

Stabilization

Racking helps stabilize the mead by removing excess sediment that could otherwise lead to spoilage or off-flavors if left in contact with the mead for too long. It also allows for the addition of stabilizers or clarifying agents, if needed.

Aging

Racking facilitates aging by transferring the mead to a new vessel where it can continue to mature and develop flavors without interference from sediment. This aging process is essential for achieving a well-rounded and harmonious final product.

Techniques and Equipment for Racking

Effective racking requires specific techniques and equipment to ensure a successful transfer and maintain the quality of the mead:

Equipment

- **Siphon or Auto-Siphon**: A siphon or auto-siphon is a common tool used for racking. It allows for a gentle transfer of mead from one vessel to another while minimizing disturbance to the sediment. Auto-siphons, with their built-in pump mechanism, are especially useful for avoiding sediment transfer.
- **Clean and Sanitize**: All equipment used for racking, including the siphon, hoses, and new fermentation vessel, must be thoroughly cleaned and sanitized to prevent contamination and spoilage.

Techniques

- **Selecting the Right Time**: Racking is typically done after primary fermentation has slowed down or completed. Signs that it is time to rack include a stable specific gravity reading and a reduction in bubbling activity. Racking can also occur during secondary fermentation if needed.
- **Minimizing Sediment Transfer**: To avoid transferring sediment, place the siphon hose slightly above the sediment layer at the bottom of the vessel. The goal is to transfer as much clear mead as possible while leaving the sediment behind.
- **Gentle Transfer**: Perform the transfer gently to minimize agitation

and disturbance of sediment. Rapid or vigorous transfers can cause sediment to mix back into the mead, affecting clarity and flavor.

Best Practices for Racking

Adhering to best practices during the racking process ensures the highest quality of the final mead:

Timing

- **Primary Fermentation**: Racking during primary fermentation is often done to separate the mead from the initial sediment and to give the yeast a cleaner environment to finish fermentation.
- **Secondary Fermentation**: Racking during secondary fermentation helps further clarify the mead and improve its overall stability and flavor. This can be done multiple times if necessary, depending on the clarity of the mead and the amount of sediment.

Sanitation

- **Cleanliness**: Ensure that all equipment involved in the racking process is thoroughly cleaned and sanitized. Contaminants can spoil the mead or introduce off-flavors.
- **Sterile Handling**: Handle the equipment with sanitized hands or gloves to avoid introducing any potential contaminants.

Monitoring

- **Gravity Readings**: Before racking, take a specific gravity reading to confirm that fermentation is complete. Racking too early can result in incomplete fermentation and potential issues with residual sugars or carbonation.

- **Visual Inspection**: Observe the clarity of the mead and the amount of sediment before racking. This helps determine the appropriate time and technique for transferring.

Additional Steps

- **Adding Finings or Stabilizers**: After racking, consider adding fining agents or stabilizers if needed to further clarify the mead and prevent potential re-fermentation or spoilage.

Impact of Racking on Mead Quality

Racking significantly impacts the quality of the final mead:

1. **Improved Clarity**: By removing sediment, racking enhances the visual clarity of the mead, resulting in a more aesthetically pleasing product.
2. **Enhanced Flavor**: Reducing contact with sediment helps eliminate off-flavors and allows the mead's true character to emerge, resulting in a cleaner and more balanced flavor profile.
3. **Extended Shelf Life**: Proper racking contributes to the overall stability and longevity of the mead, helping to prevent spoilage and maintain quality over time.
4. **Aging Potential**: Racking supports aging by providing a cleaner environment for the mead to mature, allowing for better flavor development and integration.

Aging is a critical stage in mead making that enhances the complexity and refinement of the final product. The transfer of mead for aging involves moving the mead from the primary fermentation vessel to a secondary vessel where it will mature. Knowing when and how to transfer mead for aging ensures that the mead develops optimal flavors and clarity. This essay provides a concise overview of the timing and techniques involved in transferring mead for aging.

When to Transfer Mead for Aging

The timing of the transfer to aging is crucial for achieving the best results:

Completion of Primary Fermentation

Mead should be transferred to the aging vessel after primary fermentation has slowed significantly or is complete. This is typically indicated by a stable specific gravity reading over a few days and a reduction in bubbling activity in the airlock. The yeast should have largely completed its work of converting sugars into alcohol.

Clear Signs of Sediment

By the end of primary fermentation, a significant amount of sediment, or lees, will have settled at the bottom of the fermentation vessel. Transferring the mead before excessive sediment accumulates helps prevent off-flavors and maintains clarity.

Aging Goals

Consider the style of mead and desired aging duration. Mead with higher alcohol content or certain styles may benefit from extended aging, while others may reach their peak sooner. The timing of the transfer should align with the intended aging period and the mead maker's goals.

CHAPTER SIX

How to Transfer Mead for Aging

Proper techniques for transferring mead ensure that the aging process enhances rather than hinders the final product:

Sanitize Equipment

Before transferring, thoroughly clean and sanitize all equipment, including the siphon or auto-siphon, hoses, and the new fermentation vessel. This prevents contamination and spoilage during the transfer.

Prepare the New Vessel

Select an appropriate aging vessel, such as a carboy or glass demijohn, that is clean and sanitized. The vessel should be large enough to accommodate the mead with some headspace to allow for any additional expansion or settling.

Minimize Sediment Transfer

Use a siphon or auto-siphon to carefully transfer the mead from the primary vessel to the aging vessel. Place the siphon slightly above the sediment layer to avoid disturbing it. Gentle, slow movement helps prevent sediment from mixing back into the mead.

Seal the Vessel

After transferring, seal the aging vessel with an airlock or a tightly fitted cap to minimize oxygen exposure and prevent contamination. Oxygen can lead to oxidation, which negatively affects flavor and stability.

Monitor and Age

Once transferred, monitor the mead for any signs of continued fermentation or other issues. Store the aging vessel in a cool, dark place with a stable temperature to promote optimal aging. Allow the mead to age for the desired period, which can range from several weeks to several months, depending on the mead style and goals.

7

CHAPTER SEVEN

AGING AND MATURATION

Aging, I discovered, is not merely a waiting period but a transformative phase. It's during this time that mead matures and develops complexity. When I compare a bottle of mead that has aged for a few months with one that's just been bottled, the difference is remarkable. Young mead, while still enjoyable, often exhibits sharp, unrefined flavors. The honey's sweetness can be overwhelming, and the overall profile may lack depth.

As I experimented with aging, I began to appreciate how the flavors meld and evolve over time. Aging allows the mead to mellow out, with harsh notes softening and the various flavors integrating into a harmonious blend. This maturation process is akin to the way a fine wine or whiskey improves with age. The mead develops a nuanced character that I wouldn't have achieved without letting it rest. The honey's sweetness becomes more balanced, the yeast's byproducts integrate smoothly, and any additional ingredients, like fruits or spices, meld into a cohesive symphony.

The importance of aging also extends to the clarity and texture of the mead. Young mead can be cloudy and have a rough mouthfeel. Aging helps settle out sediments and improves the overall mouthfeel, leading to a smoother, more polished final product. I've learned that the extra time allows for natural clarification, reducing the need for additional filtering or fining agents.

Perhaps one of the most intriguing aspects of aging is its role in flavor development. When I add fruits, spices, or herbs to my meads, aging gives these additional ingredients time to impart their flavors fully. This process can turn a straightforward mead into a complex and layered experience. The aging period is a chance for creativity to shine, as I can experiment with different ingredients and aging times to achieve the desired flavor profile.

In my experience, aging is not a one-size-fits-all process. Each batch of mead may require a different aging timeline, depending on the ingredients and style. While some meads are delightful fresh, others reveal their true potential only after extended aging. I've learned to keep meticulous notes on each batch, tracking how aging impacts the flavor and quality over time. This practice has been invaluable in fine-tuning my mead-making skills and achieving consistently excellent results.

In the end, aging is an art form in its own right. It requires patience, experimentation, and an appreciation for the gradual evolution of flavors. The process has taught me that while mead making starts with a recipe, it's the aging that transforms a simple brew into a masterpiece. Each bottle that emerges from the aging process is a testament to the journey from raw ingredients to a refined, exceptional beverage. So, as I continue my mead-making adventures, I embrace aging as not just a step in the process, but as a crucial element that defines the quality and character of my mead.

The Importance of Aging in Mead

Aging is a concept often associated with fine wines, spirits, and artisanal cheeses, but its importance extends far beyond these well-known examples.

CHAPTER SEVEN

Aging is a transformative process that affects a wide range of products and experiences, from fermented beverages like mead to human wisdom. Understanding the significance of aging reveals its role in enhancing quality, complexity, and depth.

To appreciate aging, we must first recognize that it involves more than mere passage of time. Aging is a deliberate process that influences how substances or entities evolve. Take, for instance, the aging of mead. Mead, often referred to as honey wine, undergoes significant changes during its maturation. Freshly brewed mead can be overly sweet or sharp, but with time, the flavors meld, and the harsh edges soften. Aging allows the components—honey, yeast, and any additional ingredients like fruits or spices—to integrate fully, resulting in a more balanced and complex flavor profile. This transformation underscores the value of patience in the pursuit of perfection.

Similarly, in the realm of fine spirits, aging plays a crucial role. Whiskey, for example, develops its distinctive character through time spent in oak barrels. During this period, the spirit interacts with the wood, absorbing flavors and undergoing chemical changes that enhance its complexity and smoothness. The process also helps mellow out any initial harshness, leading to a more refined product. The same principles apply to other spirits, such as rum and brandy, where aging imparts depth and sophistication.

Aging is not confined to consumables. It is also a vital aspect of human development and experience. As individuals age, they accumulate knowledge, skills, and wisdom that shape their perspectives and abilities. This personal growth is akin to the maturation process in products; it involves a gradual refinement and deepening of qualities that contribute to greater insight and understanding. Just as aged products often achieve a superior state, individuals too can attain a more nuanced and enriched outlook on life through their experiences.

In a broader context, aging has implications for various fields, from art to technology. In art, for example, the passage of time can enhance the value and appreciation of a work. Paintings and sculptures may gain historical significance and aesthetic appeal as they age, making them more coveted and revered. Similarly, in technology, some innovations become more valuable over time as they become more refined, adapted, or integrated into broader systems.

One of the key elements of aging is its ability to reveal hidden potential. Many products and experiences may not show their full promise immediately. Aging provides the time needed for latent qualities to emerge and develop. It is a process that allows for the slow and steady accumulation of value, whether in the form of refined flavors, deepened understanding, or enriched experiences.

However, it is important to note that aging is not always beneficial. The outcome depends on various factors, including the initial quality of the subject and the conditions under which it ages. For instance, not all meads or wines improve indefinitely with age; some may reach an optimal point and then decline. Similarly, personal growth requires active engagement and reflection; mere passage of time without introspection may not lead to meaningful development.

Aging Vessels

Aging vessels are integral to the maturation process of many products, particularly in the realms of beverage production, such as winemaking and spirit distillation. These vessels—ranging from oak barrels to stainless steel tanks—play a crucial role in shaping the final product by influencing its flavor, texture, and overall quality. Understanding the nuances of aging vessels reveals their profound impact on the maturation process.

CHAPTER SEVEN

Types of Aging Vessels

1. Oak Barrels

Oak barrels are perhaps the most iconic aging vessels in the world of spirits and wine. They have been used for centuries due to their unique properties and the way they interact with their contents. Oak barrels come in various sizes and types, including American oak, French oak, and Hungarian oak. Each type imparts different characteristics to the product. For instance, American oak tends to give a stronger vanilla and coconut flavor, while French oak contributes more subtle, spicy notes.

The interaction between the liquid and the oak occurs in several ways:

- **Extraction:** The liquid extracts compounds from the oak, such as tannins, lignins, and vanillin, which contribute to the flavor profile and mouthfeel.
- **Oxidation:** The porous nature of oak allows for slow oxygen exchange, which helps soften the flavors and reduce harshness.
- **Evaporation:** Aging in oak barrels also leads to evaporation of water and alcohol, concentrating the flavors and increasing the complexity of the product.

2. Stainless Steel Tanks

Stainless steel tanks are commonly used in the aging of wines and spirits where the primary goal is to preserve the freshness and purity of the product. Unlike oak barrels, stainless steel is inert and does not impart any flavors. This makes it ideal for products where a clean, unadulterated flavor is desired. Additionally, stainless steel tanks are easier to clean and maintain, and they do not suffer from issues like leaks or contamination.

3. Concrete Vessels

Concrete vessels, including the popular "egg" shape, are a less conventional choice but are valued for their unique characteristics. Concrete is porous, allowing for some micro-oxygenation, but it does not impart the same flavors as oak. It helps to regulate temperature and can provide a stable environment for aging. This can be particularly useful for wines that benefit from a more controlled maturation process.

4. Glass Carboys

Glass carboys are often used in home brewing and small-scale winemaking. They are valued for their non-reactive properties and the ability to easily observe the aging process. While they do not contribute to flavor development like oak barrels, they are useful for aging products that benefit from extended maturation without additional flavor influence.

Impact of Aging Vessels on Maturation

1. Flavor Development

The choice of aging vessel significantly impacts the flavor profile of the final product. Oak barrels, for instance, impart a range of flavors from vanilla and caramel to spice and smokiness, depending on the type of oak and the level of toasting or charring. The interaction between the liquid and the wood is crucial for developing these complex flavors.

In contrast, stainless steel tanks preserve the original characteristics of the liquid, which is ideal for varietal wines or spirits where the purity of the base ingredients is paramount. Concrete vessels offer a middle ground, providing some textural changes without overwhelming the product with additional flavors.

2. Texture and Mouthfeel

Aging vessels also influence the texture and mouthfeel of the product. Oak barrels can contribute to a smoother, fuller mouthfeel due to the extraction of compounds that interact with the liquid. Stainless steel tanks maintain a cleaner, crisper texture, while concrete vessels can offer a slightly rounded mouthfeel due to their unique interaction with the product.

3. Oxidation and Maturation

The rate and extent of oxidation are critical factors in the aging process. Oak barrels allow for gradual oxygen exchange, which can mellow harsh flavors and integrate complex notes. Stainless steel tanks, being airtight, prevent oxidation, preserving the freshness of the product. Concrete vessels offer a controlled level of oxidation, balancing between the extremes of oak and stainless steel.

Aging vessels are more than mere containers; they are essential tools in the maturation process that influence the final character of wines, spirits, and other beverages. Each type of vessel—be it oak barrels, stainless steel tanks, concrete vessels, or glass carboys—offers distinct advantages and impacts the aging process in unique ways. By understanding the roles and effects of different aging vessels, producers can harness their properties to craft products of exceptional quality and complexity. The choice of vessel, therefore, is a critical decision that shapes not only the flavor but the overall experience of the final product.

Monitoring and Tasting During Aging

The aging process in mead-making is crucial for developing complex flavors and achieving a refined final product. However, aging is not a passive stage;

it requires active involvement through careful monitoring and tasting to ensure that the mead matures optimally.

The Importance of Monitoring During Aging

Monitoring is essential during the aging process for several reasons:

1. Quality Control

Regular monitoring helps identify any issues early, such as off-flavors, unwanted microbial growth, or contamination. This proactive approach allows for timely interventions to rectify problems and ensure the mead's quality is maintained.

2. Progress Assessment

Monitoring helps track the progress of aging, providing insight into how the mead is evolving. This information is vital for deciding when to undertake additional steps, such as racking, filtering, or adding fining agents.

3. Adjustment and Optimization

Through monitoring, mead-makers can adjust aging conditions to optimize the final product. This might involve controlling temperature, humidity, or adjusting the aging duration based on the observed changes.

4. Consistency

For commercial mead-makers, consistent quality is essential. Monitoring ensures that each batch adheres to the desired standards, leading to a consistent and reliable product.

Monitoring Methods

Effective monitoring involves several techniques:

1. **Visual Inspection**

Regularly check the mead for clarity, color changes, and the presence of sediment. Cloudiness or unusual colors can indicate issues that need addressing.

2. **Sampling**

Periodically take small samples to assess flavor development. This practice helps gauge how the mead is progressing and whether it's meeting the desired flavor profile.

3. **Temperature Control**

Maintain and monitor the aging environment's temperature, as fluctuations can impact the aging process. Consistent temperatures help ensure uniform maturation.

4. **Hydrometer Readings**

Use a hydrometer to monitor the specific gravity of the mead. This can provide information about residual sugar levels and fermentation progress.

The Role of Tasting During Aging

Tasting is a critical component of the aging process, providing insights into flavor development and the overall quality of the mead. Here's why tasting is important:

1. **Flavor Development**

Tasting helps assess how the flavors are evolving over time. It allows meadmakers to determine if the mead is developing as intended and to make adjustments if necessary.

2. Identifying Off-Flavors

Regular tasting can reveal off-flavors or undesirable characteristics that might not be evident through other monitoring methods. Early detection of these issues allows for corrective measures.

3. Deciding on Further Steps

Tasting informs decisions about whether additional aging is needed or if interventions such as blending, adjusting sweetness, or adding other ingredients are required.

4. Evaluating Maturation

Tasting provides an overall assessment of the mead's maturation. It helps determine if the mead has reached its optimal flavor profile or if further aging will enhance its qualities.

Best Practices for Monitoring and Tasting

To maximize the effectiveness of monitoring and tasting during aging, consider the following best practices:

1. Consistency

Implement a regular schedule for monitoring and tasting. Consistent intervals ensure that you track the mead's development accurately and make informed decisions.

2. **Cleanliness**

Maintain strict hygiene when handling the mead and sampling. Use sanitized equipment to prevent contamination and preserve the integrity of the mead.

3. **Documentation**

Keep detailed records of observations, tasting notes, and any adjustments made. This documentation helps track the aging process and informs future mead-making endeavors.

4. **Sensory Evaluation**

Use a structured approach to tasting, evaluating factors such as aroma, flavor, mouthfeel, and aftertaste. Compare the mead to its intended flavor profile and assess any deviations.

5. **Temperature and Storage**

Ensure that the aging environment is stable and controlled. Proper storage conditions, such as avoiding direct sunlight and maintaining appropriate humidity levels, contribute to successful aging.

Adjusting Flavors Post-Fermentation

When making mead, adjusting flavors post-fermentation becomes one of the most crucial steps to perfecting your final product. In my experience, fermentation doesn't always produce the exact balance of flavors you're aiming for, especially with all the unpredictable nuances that yeast and fermentation conditions can throw into the mix. That's where flavor adjustments come in, turning a good mead into a great one.

Understanding the Baseline

Once fermentation is complete, the first step is always tasting the mead. I like to pour a small sample into a glass, letting it sit for a minute to open up. Just like wine, mead can have an evolving flavor profile, so giving it a moment to breathe can reveal more of what's happening inside the bottle. At this stage, I try to focus on the fundamentals: sweetness, acidity, bitterness, mouthfeel, and aroma.

The taste right after fermentation may not be a true reflection of the final product. I've often found my meads still "green" at this point. The flavors can be harsh, and the alcohol might feel too prominent. This is normal, but even with those rough edges, I start noting which flavors are already in balance and which ones need tweaking.

Adjusting Sweetness

One of the most common adjustments I make is to the sweetness level. Yeast sometimes ferments all of the sugars, leaving the mead drier than I intended. To fix this, I prefer backsweetening, which involves adding more honey or another sweetener post-fermentation.

The key to backsweetening is stabilizing the mead first. If I don't do this, adding more sugar will just kickstart fermentation again. To avoid that, I typically add potassium sorbate and potassium metabisulfite. These act as yeast inhibitors, ensuring the yeast won't go after the new sugars I introduce. After stabilization, I carefully dissolve some honey in a bit of warm water and gradually add it to the batch, tasting frequently until I hit the desired level of sweetness.

The honey I use for backsweetening isn't always the same type I used for fermentation. Sometimes, I'll experiment with a more robust honey like buckwheat or wildflower to add depth. The trick is to balance the sweetness without overwhelming the other flavors.

Balancing Acidity

If the mead feels flat or dull, adding acidity can brighten it up. Mead naturally lacks a lot of the acids found in wine or cider, so it sometimes needs a little help. I typically use citric acid or tartaric acid, but you could also add lemon juice for a more natural option.

The key here is moderation. Too much acid can turn a well-balanced mead into something sour and unappealing. I've found that adding small amounts, dissolving them in water, and tasting after each addition works best. Usually, the acidity adjustment is more subtle—enough to bring the mead into balance without making it tart.

Adjusting Bitterness and Tannins

Sometimes the mead turns out too bitter, usually from ingredients like hops (in the case of a braggot) or certain fruit skins that introduce tannins. Tannins add structure to a mead, but too much can make the drink astringent and harsh.

If the bitterness is overpowering, I dilute it with water or add more honey to counterbalance it. In extreme cases, I've found that blending the batch with another, less tannic mead can help soften the bitter edge. Oak aging also works wonders for mellowing out overly tannic or bitter meads, as the slow infusion of oak compounds can smooth the rough edges.

For meads that lack tannin structure altogether, I sometimes add tannin powder or even brew a strong tea (black tea works best) and incorporate that into the batch. It's another balancing act—too little and the mead feels weak, too much and it's like sucking on a tea bag. I aim for a level that adds depth without dominating the flavor.

Enhancing Flavor with Spices and Fruits

When I want to take my mead's flavor to the next level, I sometimes infuse it with additional spices or fruits. A cinnamon stick, vanilla bean, or even fresh fruit can dramatically change the character of a mead after fermentation.

The trick is knowing when to stop. Over-infusing can lead to overpowering, unbalanced flavors. So, when I add something like cinnamon, I keep a close eye on it, tasting every day until it reaches the desired intensity. Vanilla, on the other hand, can take a few weeks to mellow and integrate into the mead. Patience is key, as flavors tend to evolve during infusion.

Fruits offer a lot of flexibility too. I've added fresh raspberries or cherries directly into the secondary fermenter to bring a burst of fresh flavor to a fruit mead. Frozen fruits work well because freezing breaks down the cell walls, releasing more juice into the mead. However, this can introduce extra sweetness, acidity, and even tannins, so I'm always mindful of how each element will shift the overall balance.

Alcohol Content and Dilution

Occasionally, the alcohol content of my mead feels too high, giving the drink a hot, burning sensation that masks the other flavors. While aging will usually smooth this out, sometimes I dilute the mead with a bit of distilled water or even a light mead I have on hand. This can soften the alcohol intensity while bringing the flavors into better harmony.

When I dilute, I always make sure to adjust the other elements like sweetness and acidity. Adding water can flatten both, so I may need to add more honey or acid after dilution.

Clarifying and Final Touches

A mead's appearance can influence how we perceive its flavor. Hazy or cloudy mead can taste "unfinished" or "raw," so clarity is often the final step. After all my flavor adjustments, I let the mead settle and clear on its own, but if it's stubbornly cloudy, I might use a fining agent like bentonite or sparkolloid.

Clarifying agents can strip some flavors from the mead, so it's important to re-taste afterward to see if anything needs a final tweak. In my experience, the changes are subtle, but there's always a possibility of needing to adjust sweetness or acidity again.

Bottling and Aging

Once everything is in balance, I bottle the mead, but the journey doesn't end there. Aging is where the flavors truly come together. Over time, the alcohol mellows, the honey integrates more fully, and the overall experience becomes smoother. Some of my best meads took over a year to really hit their stride, proving that patience is just as important as skill when making great mead.

Clarifying Mead

Clarifying mead is a pivotal step in the mead-making process, aimed at ensuring a clear, aesthetically pleasing final product. This process involves removing unwanted particles and compounds from mead that can affect its appearance, taste, and stability. A well-clarified mead not only looks more appealing but also tends to have a smoother mouthfeel and better flavor clarity. In this essay, we will delve into the importance of clarification, explore the various methods used, and discuss best practices for achieving a perfectly clear mead.

The Importance of Clarification

Clarification is essential for several reasons:

Aesthetic Appeal

Clear mead is visually more appealing than a cloudy or hazy one. The clarity of the final product reflects the care and precision of the mead-making process.

Flavor Profile

Particles such as yeast, proteins, and polyphenols can impart off-flavors or unpleasant textures. Clarifying mead helps remove these elements, resulting in a cleaner, more refined taste.

Stability

Unwanted particles can contribute to potential spoilage or spoilage. By removing these particles, you enhance the stability of the mead, reducing the risk of off-flavors or spoilage over time.

Mouthfeel

Clear mead generally has a smoother mouthfeel. The presence of sediment or particles can create a gritty or unpleasant texture.

Methods of Clarification

Several methods can be employed to clarify mead, each with its own set of advantages and applications. These methods can be broadly categorized into natural and mechanical techniques, as well as the use of fining agents.

1. Natural Clarification

Natural clarification relies on time and gravity to settle out particles. This method involves allowing the mead to rest undisturbed in a container, typically in a cool, dark environment. As the mead ages, sediment and particles gradually settle to the bottom. This method is often used in conjunction with other techniques to achieve optimal clarity.

2. Mechanical Clarification

Mechanical clarification involves using physical means to separate particles from the mead. Common methods include:

- **Racking**: Racking involves transferring the mead from one vessel to another, leaving sediment behind. This is often done multiple times during the aging process to gradually remove sediment.
- **Filtration**: Filtration uses various types of filters to remove particles from the mead. Options include:
- **Paper Filters**: These are used to remove larger particles and are suitable for initial clarification.
- **Plate Filters**: These provide finer filtration and can remove smaller particles, resulting in a clearer product.
- **Membrane Filters**: These offer the highest level of filtration and are capable of removing even the smallest particles.

3. Fining Agents

Fining agents are substances added to the mead to aid in the clarification process. They work by binding to particles and causing them to clump together, making it easier for them to settle out. Common fining agents include:

- **Bentonite**: A clay-like substance that is effective at removing proteins and haze-forming compounds. Bentonite is particularly useful in dealing with protein haze.
- **Gelatin**: A natural protein derived from animal collagen, gelatin helps remove yeast and other particles. It is often used in combination with other fining agents for optimal results.
- **Isinglass**: Derived from fish bladders, isinglass is effective at removing

yeast and other particulate matter. It is often used in conjunction with other fining agents.
- **Activated Charcoal**: Also known as activated carbon, this fining agent is used to remove off-flavors and odors. It is less commonly used for physical clarification but can improve the overall sensory profile of the mead.

Best Practices for Clarifying Mead

To achieve the best results in clarifying mead, consider the following best practices:

1. **Patience**: Natural clarification takes time, and rushing the process can lead to incomplete results. Allow the mead ample time to settle before proceeding with further clarification steps.
2. **Cleanliness**: Maintain strict cleanliness throughout the clarification process to avoid contamination. Ensure that all equipment, including filters and vessels, is thoroughly sanitized.
3. **Monitoring**: Regularly monitor the mead during the clarification process to assess progress and determine when additional clarification methods may be needed.
4. **Combination of Methods**: Often, a combination of natural, mechanical, and fining techniques yields the best results. For example, using racking and filtration in conjunction with fining agents can achieve optimal clarity.
5. **Testing**: Before bottling, test the clarity of the mead to ensure it meets your standards. If necessary, perform additional clarification steps to achieve the desired clarity.

III

STORAGE AND ADVANCED MEAD MAKING TECHNIQUES

8

CHAPTER EIGHT

BOTTLING AND STORAGE

When it comes to bottling and storing mead, I always get a sense of accomplishment, like putting the final touches on a masterpiece. After all the hard work—from selecting the right honey, yeast, and fermentables, to adjusting flavors post-fermentation—it's finally time to bottle this liquid gold.

First, I make sure my mead has finished fermenting. One way to tell is when the airlock stops bubbling for several days, but I prefer checking the specific gravity with a hydrometer. If it's stable over a few readings, I know the yeast has done its job and I'm ready to proceed. I also take a sip at this stage, just to make sure it tastes right. There's something so satisfying about knowing the flavor is spot on before it gets bottled.

Now, sterilization is key. The last thing I want is to ruin all those months of fermentation by being careless with sanitation. I clean and sanitize every bottle, stopper, siphon, and hose. I usually use glass bottles because they preserve the mead better, and honestly, there's something nostalgic about popping a cork from a glass bottle years later.

Siphoning the mead into bottles feels like the final step in this long, rewarding process. I take care not to disturb the sediment at the bottom of the carboy as I siphon, aiming for a crystal-clear pour. I leave about an inch of headspace at the top of each bottle, which allows room for any gas that might be released during storage.

Then comes the choice between corks or screw caps. I prefer corks for long-term aging, but screw caps are great for convenience if I plan on drinking the mead within a year or two. After sealing the bottles, I label them with the date and any other information I want to remember—like the type of honey I used or any flavor adjustments I made. Trust me, you'll want to recall those details when you pop open a bottle in a few years.

Storage is the next big consideration. Mead ages beautifully, so I store my bottles in a cool, dark place like a cellar. Ideally, the temperature stays between 50 and 60°F. Patience is key here; while some meads are drinkable within a few months, I find that waiting at least six months to a year brings out deeper, more complex flavors. For certain batches, I'll wait even longer, knowing that time can work wonders on the taste.

Choosing Bottles and Closures

When it comes to mead making, every detail matters, from the honey selected to the temperature of fermentation. However, an often overlooked yet crucial part of the process is the selection of bottles and closures. This final step can significantly impact not only the presentation of your mead but also its longevity, flavor development, and ease of use. The choice of bottles and closures involves a delicate balance between functionality and tradition, with options ranging from practical screw caps to the more classical appeal of corks.

Bottle Types

CHAPTER EIGHT

The first decision to make when bottling mead is selecting the type of bottle. While the beverage can be bottled in various containers, certain bottle types offer distinct advantages based on the mead's intended use and aging process.

Glass Bottles

Glass bottles are the go-to choice for mead makers, largely due to their non-reactive nature. Glass does not interact with the mead, ensuring that the flavors remain untainted during storage. Within the glass bottle category, there are various shapes and styles to choose from, each with its own characteristics and traditions.

- **Wine Bottles**: Wine-style bottles, often seen in 750 mL or 375 mL sizes, are a popular option for mead. These bottles are elegant, timeless, and work well with both corks and screw caps. Wine bottles come in a variety of colors, but darker shades like amber or green are ideal for protecting the mead from harmful light exposure, which can degrade its quality over time.
- **Beer Bottles**: For meads that are carbonated, such as sparkling meads, beer bottles are often a better choice. These bottles are built to withstand pressure and are usually capped with crown caps, which provide an airtight seal necessary for maintaining carbonation.
- **Flip-Top Bottles**: These bottles come with built-in closures, featuring a ceramic stopper and a rubber gasket that can be secured with a wire bail. Flip-top bottles are convenient and offer a rustic, handcrafted appearance, making them ideal for small batches or gift-giving.

Plastic Bottles

Though not traditionally used in mead making, plastic bottles can be useful for short-term storage or sampling. The primary drawback of plastic is its permeability to oxygen, which can lead to oxidation of the mead over time.

Plastic also does not provide the same aesthetic appeal or longevity as glass, making it less suitable for meads intended for aging.

Choosing the Right Closure

Once the bottle type has been selected, the next critical decision is the type of closure. The closure not only affects the aging process but also plays a role in the ease of opening the bottle and how the mead is perceived by consumers.

CORKS

For centuries, corks have been the closure of choice for wine and mead. They provide an air-tight seal while still allowing a small amount of oxygen exchange, which can enhance the aging process for certain types of mead. However, corking comes with its own set of challenges and considerations.

- **Natural Corks**: Made from the bark of the cork oak tree, natural corks offer a traditional look and feel. They allow minimal oxygen transfer, which can help meads evolve over time. However, natural corks are susceptible to "cork taint," a phenomenon caused by a compound called TCA (trichloroanisole) that can spoil the mead.
- **Synthetic Corks**: As an alternative to natural cork, synthetic corks provide the same air-tight seal without the risk of cork taint. They are also easier to work with and more consistent in quality. However, they may not provide the same degree of oxygen exchange as natural corks, which could impact the aging process of mead.
- **Agglomerated Corks**: Made from compressed cork particles, agglomerated corks are less expensive than natural corks and are often used for meads intended for shorter aging periods. They do not offer the same longevity or premium feel as natural or synthetic corks but are a viable option for everyday meads.

CHAPTER EIGHT

SCREW CAPS

Screw caps have gained popularity in recent years, especially for wines and meads that are meant to be consumed within a shorter time frame. Screw caps provide a perfect seal, preventing any oxygen from entering the bottle. This makes them ideal for meads that are not intended for long-term aging, as they preserve the original flavors and aromas without allowing for further development in the bottle.

One of the significant advantages of screw caps is their ease of use. They eliminate the need for corkscrews and allow for resealing the bottle if it is not consumed in one sitting. This convenience makes screw caps an appealing option for meads that are enjoyed more casually or shared in a social setting.

CROWN CAPS

For sparkling meads or those that have been carbonated, crown caps are often the closure of choice. Similar to the caps used on beer bottles, crown caps provide an airtight seal that can withstand the pressure created by carbonation. They are easy to apply and remove, though they require a bottle capper for sealing.

Balancing Aesthetics and Practicality

Ultimately, the choice of bottles and closures comes down to a combination of aesthetics, practicality, and the intended use of the mead. For meads that are intended to be aged for years, corks and wine bottles provide the elegance and functionality needed for long-term storage. On the other hand, for meads that will be consumed more quickly, screw caps or flip-top bottles offer a convenient and effective solution.

The appearance of the bottle also plays a role in how the mead is perceived. A well-chosen bottle can elevate the presentation, making the mead feel like a premium product even before the first sip. Whether it's the rustic charm of a flip-top bottle or the sophisticated elegance of a corked wine bottle, the right choice can set the tone for the entire mead-drinking experience.

Proper Bottling Techniques

Proper bottling techniques are essential to avoid spoilage, oxidation, and to ensure that the mead ages gracefully. Here are steps to take when bottling;

1. Preparation and Planning

Before starting the bottling process, it is crucial to have everything organized and prepared. This involves setting up a clean and efficient workspace where you can comfortably and systematically work through the bottling process.

- **Clean Work Area**: Choose a clean, well-lit area to work. Lay down a sanitized table cover or disposable plastic sheet to protect surfaces and catch any spills.
- **Gather Equipment**: Ensure you have all necessary equipment ready, including bottles, a siphon or racking cane, tubing, a funnel, and any additional items like labels or a bottle capper if applicable. Make sure that everything is sanitized.

2. Determining Readiness for Bottling

Before bottling, confirm that the mead is fully fermented and ready to be bottled.

- **Hydrometer Testing**: Use a sanitized thief or pipette to draw a sample

of mead and measure its specific gravity with a hydrometer. The gravity readings should be stable over several days to ensure fermentation has completed.
- **Tasting**: Conduct a taste test to ensure the mead has reached the desired flavor profile. This is the last chance to make any final adjustments before bottling.

3. Siphoning the Mead

Siphoning is a delicate process that requires care to avoid disturbing sediment and introducing oxygen into the mead.

- **Setup**: Place the fermenter on a higher surface, such as a table or countertop, allowing gravity to assist with the transfer.
- **Insert Siphon**: Insert the siphon or racking cane into the fermenter, positioning it above the sediment at the bottom. Ensure that the tubing is sanitized and attached securely.
- **Start Siphoning**: Begin siphoning by either manually pumping the auto-siphon or by starting the flow with a gravity siphon. Keep the end of the tubing submerged in the mead to minimize splashing and oxygen exposure.
- **Fill Bottles**: Insert the tubing into the bottles and gently transfer the mead. Fill each bottle slowly to avoid splashing and to reduce the risk of oxidation. Leave approximately 1 to 1.5 inches of headspace at the top of each bottle to accommodate any gas expansion or sediment settling.

4. Handling Bottles

As each bottle is filled, handle them carefully to ensure they remain clean and free from contaminants.

- **Avoid Touching the Inside**: Do not touch the inside of the bottles or the bottle neck to prevent contamination. If you need to adjust the positioning of the bottle, use sanitized tools.
- **Check for Sediment**: As you approach the end of the siphoning process, monitor the mead flow for any sediment being drawn into the tubing. Stop siphoning before sediment reaches the bottle.

5. Sealing the Bottles

The closure of the bottle is critical for preserving the quality of the mead and preventing spoilage.

- **Apply Closures**: If using corks, make sure they are properly inserted using a corker, ensuring a tight fit without damaging the cork. For screw caps, align and secure them tightly to prevent any leakage. If using crown caps, use a capper to seal them firmly onto the bottle.
- **Ensure a Good Seal**: Double-check that all closures are properly sealed. Any gaps or loose closures can lead to contamination or spoilage.

6. Labeling

Proper labeling is not only a professional touch but also a practical way to track and organize your mead.

- **Prepare Labels**: Create and print labels with important information such as the date of bottling, the type of mead, alcohol content, and any specific notes about the flavor or ingredients.
- **Affix Labels**: Apply the labels to the bottles once they are sealed. Ensure that the labels are placed on a clean, dry surface of the bottle to ensure adhesion.

7. Storage

Proper storage is essential for maintaining the quality of the mead during aging.

- **Store in a Cool, Dark Place**: Place the bottles in a cool, dark area where the temperature remains stable, ideally between 50-60°F. Light and temperature fluctuations can negatively impact the mead's flavor and aging process.
- **Position Bottles Correctly**: Store bottles horizontally if using corks to keep the corks moist and ensure a tight seal. For screw-capped or crown-capped bottles, storing upright is acceptable.

8. Aging and Monitoring

After bottling, patience is key. The aging process will allow the mead to develop complexity and depth of flavor.

- **Monitor Aging**: Regularly check the storage conditions and the bottles for any signs of leakage or spoilage.
- **Taste Testing**: Over time, sample the mead to track its development. This will help gauge its readiness for consumption and determine the best time to enjoy it.

Proper bottling techniques are essential to preserving the quality and integrity of mead. By carefully managing each step—from preparation and siphoning to sealing and storage—you ensure that the mead remains uncontaminated and develops its full potential. The attention to detail during bottling reflects the effort invested throughout the mead-making process, culminating in a final product that can be enjoyed for years to come.

Labeling and Record-Keeping

When it comes to bottling mead, labeling and record-keeping are more than just formalities—they're essential practices that help me track the evolution of my mead, manage my inventory, and ensure that I can replicate successful batches or learn from less successful ones.

LABELING

For me, labeling is both a practical necessity and a chance to add a touch of personality to each bottle. Here's how I approach it:

1. Choosing Label Materials

I prefer using durable, waterproof labels that can withstand any accidental spills or moisture. My go-to is often glossy or matte adhesive labels that stick firmly to the bottles and look professional. I also use a label maker or a computer to design and print my labels, ensuring they're clear and legible.

2. Designing the Labels

I design my labels with both functionality and aesthetics in mind. Each label includes the following key information:

- **Name and Type of Mead**: Clearly stating whether it's a traditional mead, melomel, or any other type helps differentiate the bottles at a glance.
- **Date of Bottling**: I always include the bottling date to keep track of aging. This is crucial for monitoring how the mead develops over time.
- **Alcohol Content**: If known, I note the alcohol by volume (ABV) to give an idea of the strength of the mead.
- **Ingredients and Any Special Notes**: I list notable ingredients or

any special adjustments made, such as spices or fruits added. This information is valuable for both me and anyone who might be enjoying the mead later.
- **Tasting Notes or Descriptions**: Adding a brief description or tasting notes can be a nice touch, giving an idea of what flavors to expect and how the mead should ideally be enjoyed.

3. Applying the Labels

After the bottles are sealed and dried, I carefully apply the labels. I align them to ensure they're straight and positioned uniformly. A neat label not only looks professional but also helps in easy identification.

RECORD KEEPING

Keeping detailed records is just as important as labeling. My approach to record-keeping helps me track each batch's progress and ensure consistency across my mead-making endeavors.

1. Creating a Bottling Log

I maintain a dedicated bottling log, which can be a physical notebook or a digital spreadsheet, depending on my mood and convenience. In this log, I record:

- **Batch Details**: I write down the batch number, the type of mead, and the volume produced. This helps me track which batch is which.
- **Fermentation Data**: I include notes on the initial and final specific gravity readings, fermentation duration, and any issues encountered.
- **Ingredients and Adjustments**: Detailed records of the ingredients used, their quantities, and any adjustments made during fermentation or prior to bottling are crucial. This includes any flavor adjustments or

back-sweetening.
- **Tasting Notes**: I document my impressions of the mead at different stages—pre-bottling, post-bottling, and at various points during aging. This helps me gauge how the mead is developing over time.

2. Inventory Management

To keep track of my mead inventory, I use a simple system that logs each bottle's status. I note when a bottle is opened, when it's consumed, and any observations about its flavor profile at that time. This inventory helps manage my stock and ensures I'm aware of how much mead is aging and how much is ready to drink.

3. Review and Adjustments

Periodically, I review my records and labels to assess trends and outcomes. If a particular batch turns out exceptionally well, I make detailed notes on what went right, so I can replicate the success in future batches. Conversely, if a batch doesn't turn out as expected, I analyze my records to identify any potential issues or areas for improvement.

Labeling and record-keeping are integral to my mead-making process. They not only help me stay organized but also enhance the quality and enjoyment of my mead. By meticulously labeling each bottle and maintaining comprehensive records, I ensure that I can track my progress, learn from each batch, and continue to refine my mead-making skills. The combination of detailed labeling and diligent record-keeping transforms the art of bottling from a mere final step into a crucial aspect of the mead-making journey.

9

CHAPTER NINE

EXPERIMENTING WITH FLAVORS

Experimenting with flavors in mead making is one of the most exciting aspects of crafting this ancient beverage. For me, it starts with an idea—perhaps inspired by a favorite fruit, spice, or even a cocktail recipe. I'll often spend time researching different flavor combinations, considering how various ingredients will interact with the honey and the yeast.

I begin by brewing a basic mead recipe, which serves as my blank canvas. This foundational mead, while simple, gives me a solid base to which I can add different elements later on. Once it's fermented and has had time to mature, I'll start experimenting with different flavors.

For fruit flavors, I might use anything from berries to tropical fruits. I find that fresh fruit works best if I want a vibrant, juicy flavor, but sometimes I'll use dried fruit or fruit extracts for a more subtle touch. Adding fruit is all about timing—too early, and the fruit flavors can ferment away; too late, and they might not meld well with the mead.

Spices and herbs are another playground for me. I love adding cinnamon, cloves, or even ginger to my mead for a warm, spicy kick. The key is to start small; spices can be overwhelming, and it's easier to add more than to take away. I'll often make a small batch with a test amount of spice and taste it before deciding to scale up.

Occasionally, I'll experiment with herbs like mint or lavender. These can be a bit more delicate, so I'll infuse them carefully, watching closely to ensure they don't overpower the mead.

Sometimes, I'll add flavorings like vanilla or cocoa nibs. These ingredients can introduce unique notes that complement the honey's natural flavors. I find that letting these infusions sit for a few days and then tasting frequently helps me achieve the perfect balance.

The final step is tasting and adjusting. I'll often blend small amounts of my flavored mead with the base mead to see how they mix. It's a bit of a science and an art, and I enjoy the process of tweaking and fine-tuning until I get the flavor just right.

Experimenting with flavors is where creativity truly shines in mead making. Each batch is a chance to discover something new, and the process of trial and error is always rewarding when I finally hit on a combination that I love.

Incorporating Fruits: Techniques and Timing

While traditional meads are simple and straightforward, modern mead makers frequently enhance their creations by incorporating fruits. This addition can add depth, complexity, and a variety of flavors to the mead. The process of incorporating fruits into mead involves careful consideration of techniques and timing to achieve the desired result. This essay delves into the methods of incorporating fruits into mead, exploring the impact on flavor, aroma, and overall quality.

CHAPTER NINE

Techniques for Incorporating Fruits

Fruit Type and Preparation

The choice of fruit can significantly influence the final product. Common fruits used in mead making include berries (such as raspberries, blueberries, and blackberries), citrus fruits (like oranges and lemons), and stone fruits (such as peaches and cherries). Each fruit offers unique flavors and aromas that can complement or contrast with the honey.

Proper preparation of fruit is crucial. Fresh fruits should be washed and peeled (if necessary), while frozen fruits should be thawed and strained to remove excess water. Fruits can be used in various forms, including whole, crushed, pureed, or juiced. The form of fruit used can affect the mead's clarity and extraction of flavors. Purees and juices tend to blend more seamlessly, while whole or crushed fruits can provide more pronounced texture and flavor.

Fruit Addition Timing

The timing of fruit addition plays a vital role in the mead-making process. There are generally three main stages where fruit can be added: primary fermentation, secondary fermentation, and during aging.

- **Primary Fermentation**: Adding fruit during primary fermentation allows the yeast to interact with the fruit, leading to a more integrated flavor profile. This technique, however, can sometimes result in a less defined fruit flavor because the fermentation process can alter and sometimes overpower delicate fruit notes.
- **Secondary Fermentation**: Adding fruit after the primary fermentation, during secondary fermentation, is a popular method. This timing ensures that the fruit flavors are preserved more effectively, as the mead has already completed the primary fermentation and is less likely to

have its flavors altered by vigorous yeast activity. The fruit is added to the carboy or secondary fermenter, and the mead is left to ferment for several weeks to several months, depending on the desired flavor intensity.

- **Aging**: Some mead makers choose to add fruit during the aging process, after fermentation has completed. This method allows for the addition of fresh fruit flavors without the risk of over-fermentation or spoilage. Fruits can be added directly to the mead, or as fruit extracts, to achieve specific flavor profiles. This technique is particularly useful for adding delicate or subtle fruit notes that might be lost if added earlier.

Fruit Processing

When adding fruit to mead, it's essential to consider the processing methods to prevent contamination and ensure the best flavor extraction. Sanitation is paramount; all equipment and fruits should be thoroughly cleaned and sanitized. Additionally, the fruit should be prepared in a way that maximizes flavor extraction. For example, fruit can be crushed or pureed to increase surface area and release more of its natural flavors.

Some mead makers use fruit extracts or concentrates as an alternative to fresh fruit. These extracts can provide consistent and potent fruit flavors without the need for extensive preparation. However, they can sometimes lack the complexity and nuance of fresh fruit.

Balancing Fruit Flavors

Balancing fruit flavors with the honey and other components of the mead is crucial. Mead makers should taste the mead periodically during the fermentation and aging process to ensure that the fruit flavors are harmonious with the honey and other ingredients. Adjustments can be made by adding more fruit, honey, or other flavorings as needed to achieve the desired balance.

Timing Considerations

The timing of fruit addition can profoundly impact the mead's final flavor profile. Adding fruit early in the fermentation process may result in a more homogenized flavor, where the fruit blends with the honey and yeast. Conversely, adding fruit later, particularly during secondary fermentation or aging, allows for more pronounced and distinct fruit flavors.

The choice of timing should align with the desired outcome. For example, if a mead maker aims for a subtle fruit undertone, adding fruit during primary fermentation might suffice. For a more robust fruit flavor, secondary fermentation or aging is preferred.

Incorporating fruits into mead making opens up a world of possibilities, allowing mead makers to experiment with flavors and create unique and personalized beverages. By carefully selecting the type of fruit, choosing the right timing for addition, and employing proper processing techniques, mead makers can enhance their meads with vibrant and complex fruit flavors. Whether aiming for a delicate fruit aroma or a bold fruit presence, understanding the nuances of fruit incorporation can elevate the quality and enjoyment of the final product.

Using Spices and Herbs

Spices and herbs can transform mead from a simple honey-based beverage into a complex and aromatic experience.

Techniques for Incorporating Spices and Herbs

Selection and Preparation

The choice of spices and herbs significantly impacts the final flavor of the mead. Commonly used spices include cinnamon, cloves, ginger, and vanilla, while herbs might include lavender, mint, rosemary, and thyme. The selection should align with the flavor profile desired for the mead.

Preparation is crucial to ensure that the spices and herbs impart their flavors effectively. Whole spices, such as cinnamon sticks or cloves, should be crushed or broken to increase their surface area. Ground spices can be added directly, but they may require more careful straining to avoid sediment. Fresh herbs should be washed and chopped, while dried herbs can be added as-is.

Infusion Methods

- **Boiling**: Spices and herbs can be added to boiling water to create a tea or infusion before being added to the mead. This method extracts flavors quickly and helps ensure that the spices and herbs are evenly distributed. The infusion should be cooled before adding it to the mead.
- **Direct Addition**: Spices and herbs can be added directly to the fermenter. This method is straightforward but requires careful monitoring to prevent over-extraction. It's often used for spices with robust flavors that can withstand direct contact with the mead.
- **Tinctures and Extracts**: For a more controlled approach, mead makers can use tinctures or extracts. Spices and herbs are steeped in alcohol to create a concentrated extract, which is then added to the mead. This method allows for precise control over the flavor intensity and minimizes sediment.
- **Secondary Fermentation**: Adding spices and herbs during secondary fermentation allows the flavors to meld without being overwhelmed by active fermentation. This technique is ideal for delicate herbs and spices that might otherwise be overpowered or altered during primary fermentation.
- **Aging**: Spices and herbs can also be added during the aging process. This approach allows the flavors to integrate slowly and develop

complexity. However, it requires careful monitoring to prevent the flavors from becoming too strong or unbalanced.

Balancing Flavors

Balancing the flavors of spices and herbs with the honey and other ingredients is key to a well-rounded mead. It's important to start with small amounts and gradually adjust based on taste. Tasting the mead periodically during fermentation and aging will help ensure that the spices and herbs contribute positively to the overall flavor profile.

Sanitation

As with any ingredient in mead making, sanitation is critical when using spices and herbs. All equipment and ingredients should be thoroughly cleaned and sanitized to prevent contamination and spoilage.

Timing Considerations

The timing of spice and herb addition is crucial to achieving the desired flavor profile. Here's a closer look at the effects of timing:

1. **Primary Fermentation**: Adding spices and herbs during primary fermentation allows for a thorough integration of flavors. However, the active fermentation process can alter the flavors, making them less distinct. This method is suitable for strong spices and herbs that can withstand the vigorous fermentation environment.
2. **Secondary Fermentation**: Adding spices and herbs during secondary fermentation provides a more controlled environment for flavor extraction. The mead has already completed most of its fermentation, allowing the spices and herbs to impart their flavors more delicately. This timing is ideal for more subtle or complex spices and herbs.

3. **Aging**: Incorporating spices and herbs during aging allows for a slow and gradual integration of flavors. This method is useful for creating nuanced and layered flavor profiles. It also offers flexibility in adjusting flavors as needed.

Examples of Spice and Herb Combinations

- **Cinnamon and Vanilla**: A classic combination that adds warmth and sweetness, perfect for spiced meads or those aiming for a dessert-like quality.
- **Ginger and Lime**: Offers a zesty and refreshing profile, ideal for lighter meads or those with tropical fruit notes.
- **Lavender and Honey**: Creates a floral and aromatic mead with a subtle sweetness, well-suited for traditional and herbaceous styles.
- **Rosemary and Orange Peel**: Adds a savory and citrusy twist, complementing meads with complex or earthy honey flavors.

Specialty Meads: Creating Metheglins, Melomels, and Others.

Specialty meads, such as metheglins, melomels, and various other variants, offer mead makers the opportunity to explore a wide range of flavors and styles. These types of mead incorporate additional ingredients beyond honey and water, leading to complex and unique beverages.

Metheglins

Definition and Characteristics

Metheglins are spiced meads, where various spices and herbs are added to enhance the flavor profile. The term "metheglin" originates from the Welsh word "meddyglyn," meaning "healing mead," as these meads were historically believed to have medicinal properties due to the herbs used.

Techniques for Creating Metheglins

Spice and Herb Selection

Choose spices and herbs based on the flavor profile you desire. Common spices include cinnamon, cloves, ginger, and cardamom. Herbs might include lavender, rosemary, thyme, or mint. The choice of spices and herbs will significantly impact the final flavor of the metheglin.

Preparation and Addition

- **Infusions**: Prepare spice or herb infusions by boiling them in water and then adding the cooled infusion to the mead. This method extracts flavors efficiently and can be adjusted to achieve the desired strength.
- **Direct Addition**: For a more robust spice presence, add spices and herbs directly to the fermenter. Crush or grind whole spices to increase flavor extraction and use a muslin bag or mesh to contain them, making it easier to remove them later.
- **Tinctures and Extracts**: Use tinctures or extracts for precise control over the spice flavors. These concentrated forms of spices or herbs are added in small amounts to achieve the desired intensity.

Timing

- **Primary Fermentation**: Adding spices and herbs during primary fermentation can lead to well-integrated flavors but may result in some loss of subtlety due to the vigorous fermentation process.
- **Secondary Fermentation**: Adding spices and herbs during secondary

fermentation allows for more controlled flavor extraction and preservation of delicate flavors.
- **Aging**: Adding spices and herbs during aging can enhance the complexity and depth of the flavors. This method requires careful monitoring to prevent over-extraction.

Melomels

Definition and Characteristics

Melomels are fruit meads, where various fruits are added to the mead. The term "melomel" comes from the Latin word "mela" (apple) and "mel" (honey), though melomels can include a wide range of fruits.

Techniques for Creating Melomels

Fruit Selection and Preparation

Choose fruits based on the desired flavor profile. Common fruits used in melomels include berries (strawberries, raspberries, blueberries), stone fruits (peaches, cherries, plums), and tropical fruits (mango, pineapple).

- **Fresh Fruits**: Wash, peel (if necessary), and prepare fresh fruits by mashing, pureeing, or cutting them into pieces.
- **Frozen Fruits**: Thaw and drain frozen fruits to remove excess water before adding them to the mead.
- **Juices and Extracts**: Fruit juices and extracts provide a more controlled way to add fruit flavors and are useful when fresh or frozen fruits are not available.

Addition and Integration

- **Primary Fermentation**: Adding fruit during primary fermentation allows the yeast to ferment the sugars from the fruit, which can lead to a more integrated fruit flavor. However, this method may also result in a less pronounced fruit character.
- **Secondary Fermentation**: Adding fruit during secondary fermentation is a popular method for preserving more distinct fruit flavors. The mead will develop a clearer fruit character and aroma, as the fermentation process is less aggressive.
- **Aging**: Fruits can also be added during the aging process for a subtle and refined fruit flavor. This approach allows the mead to develop complexity over time.

Balancing Flavors

1. Properly balance the fruit flavor with the honey to avoid overwhelming the mead. Taste the melomel periodically and make adjustments if necessary by adding more fruit or honey.

Other Specialty Meads

Cyser

Cyser is a type of mead made with apple juice or cider. It combines the characteristics of mead with the flavors of apples, resulting in a drink similar to apple cider but with a higher alcohol content.

- **Techniques**: Add apple juice or cider to the honey and water mixture before fermentation. You can use either fresh apple juice, cider, or even apple concentrates.
- **Timing**: Cyser can benefit from aging, which helps meld the apple and honey flavors. Adding spices like cinnamon or cloves can also enhance the apple flavor.

Pyment

Pyment is a mead made with grape juice or wine, combining the flavors of mead with those of grapes. It resembles a hybrid between mead and wine.

- **Techniques**: Blend grape juice or wine with honey and water. Adjust the amount of honey based on the sweetness of the grape juice or wine.
- **Timing**: Pyment can be added to the primary fermentation, or grape juice/wine can be added during secondary fermentation for more distinct wine flavors.

Braggot

Braggot is a blend of mead and beer, created by combining malted grains or beer with honey.

- **Techniques**: Brew a small batch of beer and mix it with honey and water before fermentation. Alternatively, add malt extracts or grains directly to the mead.
- **Timing**: Braggots can be brewed as a single fermentation or with separate fermentation stages for the beer and mead components, which are then blended.

10

CHAPTER TEN

SCALING UP PRODUCTION

This process not only requires a deep understanding of the fundamentals of mead making but also demands a keen grasp of business practices, resource management, and quality control.

1. Understanding the Basics of Scaling Up

Scaling up mead production fundamentally requires replicating the small-scale process on a larger scale while maintaining or enhancing quality. The basic principles of fermentation, ingredient proportions, and processing times remain consistent, but the scale introduces new variables. One of the first steps is to understand that scaling up isn't merely about increasing quantities but optimizing processes to handle the increased volume efficiently.

2. Upgrading Equipment

A critical aspect of scaling up mead production is upgrading from homebrewing equipment to industrial-scale machinery. At a small scale, mead makers often use basic fermenters, airlocks, and bottling setups. In larger-scale production, these need to be replaced or supplemented with industrial equipment, including:

- **Fermentation Tanks**: Larger fermentation tanks made from stainless steel or food-grade plastic are essential. These tanks should be equipped with temperature control systems and fermentation monitoring devices to ensure consistent fermentation conditions.
- **Heat Exchangers**: For managing the temperature of large volumes of mead, heat exchangers can help control the temperature of the wort and maintain the desired fermentation environment.
- **Filtration Systems**: Scaling up often necessitates more efficient filtration systems to clarify the mead before bottling. These systems help in removing unwanted particles and sediments that can affect flavor and clarity.
- **Bottling and Packaging Lines**: Automated bottling lines improve efficiency, speed up production, and maintain consistency. These systems can handle filling, capping, labeling, and packaging in a streamlined process.

3. Process Optimization

Scaling up production also involves optimizing various stages of the mead-making process:

- **Ingredient Handling**: At larger scales, sourcing and handling raw materials—such as honey, water, and yeast—become more complex. Establishing reliable suppliers and implementing efficient inventory management practices are crucial to maintaining consistent quality.
- **Mixing and Blending**: Larger volumes require more sophisticated mixing equipment to ensure uniformity in the mead's flavor profile.

Automated systems can help achieve precise blending and consistency across batches.
- **Fermentation Management**: Monitoring and controlling the fermentation process becomes more challenging with larger volumes. Implementing advanced sensors and control systems can help track fermentation progress and make necessary adjustments.
- **Aging and Maturation**: Scaling up may also involve expanding aging facilities. Ensuring that mead matures properly in larger barrels or tanks is critical to achieving the desired flavor profiles.

4. Quality Control

Maintaining quality at a larger scale requires robust quality control measures:

- **Standard Operating Procedures (SOPs)**: Developing detailed SOPs for every stage of production ensures consistency and quality. These procedures should cover ingredient handling, fermentation processes, and sanitation practices.
- **Testing and Analysis**: Regular testing of samples for various parameters such as alcohol content, pH levels, and microbial contamination helps in maintaining the quality of the final product.
- **Training and Personnel**: Training staff thoroughly in all aspects of mead production is vital. Skilled personnel are necessary to manage equipment, monitor processes, and ensure adherence to quality standards.

5. Market Considerations

Scaling up production also involves understanding and navigating market dynamics:

- **Regulatory Compliance**: Larger-scale operations must comply with local and national regulations regarding alcohol production, labeling, and distribution. This includes obtaining necessary permits and ensuring adherence to safety standards.
- **Distribution Channels**: Expanding production often requires establishing or enhancing distribution channels. Building relationships with distributors, retailers, and exploring direct-to-consumer sales options can help in reaching a broader market.
- **Marketing and Branding**: As production scales, branding and marketing strategies need to be adapted. Developing a strong brand identity, creating effective marketing campaigns, and leveraging social media and other platforms can help in capturing and expanding your customer base.

6. Financial and Logistical Planning

Scaling up production requires careful financial and logistical planning:

- **Budgeting and Investment**: Estimating the costs associated with upgrading equipment, expanding facilities, and managing increased production volume is essential. Securing investment or financing may be necessary to support this expansion.
- **Supply Chain Management**: Efficient supply chain management is crucial to ensuring that raw materials are consistently available and that production schedules are met.
- **Risk Management**: Identifying potential risks, such as equipment failures or supply chain disruptions, and developing contingency plans can help mitigate the impact on production.

CHAPTER TEN

Managing Fermentation in Bulk

Transitioning from small-batch to large-scale mead making has been a fascinating journey, one that required me to delve deep into the complexities of fermentation on a grand scale. The process of managing fermentation in bulk presents unique challenges and opportunities, each of which I navigated with a mix of excitement and meticulous planning. Here's a detailed account of my experience and the insights I've gained along the way.

Upgrading Equipment

My first major realization was that the equipment used for small batches simply wouldn't cut it for bulk fermentation. I quickly learned that stainless steel fermentation tanks were indispensable. These tanks, with their robust construction and sanitary properties, became the backbone of my new setup. I chose vessels with temperature control jackets and sampling ports—features that proved invaluable. Temperature control was a game-changer. I invested in systems that could maintain a consistent temperature, crucial for yeast activity and fermentation consistency.

Yeast Management

Yeast, as it turns out, is both a simple and complex organism to manage in bulk. Initially, I grappled with the right pitching rate, realizing that the same principles applied to small batches didn't directly scale up. I started using yeast starters to build up the necessary yeast cell count before adding it to the fermentation tank. This practice helped me achieve the right fermentation vigor across large volumes. I also learned the importance of selecting yeast strains that could handle the volume and conditions of bulk fermentation. It was a delicate balance, but with careful strain selection and proper propagation, I achieved more predictable results.

Monitoring and Controlling Fermentation

Managing fermentation in bulk required a new level of precision. I installed automated sensors and control systems to monitor fermentation parameters such as temperature, specific gravity, and pH. These systems allowed me to track the fermentation progress in real-time and make adjustments as needed. Proper oxygenation was another critical aspect. I used oxygenation systems to ensure the wort was well-aerated before pitching the yeast. However, managing oxygen levels throughout fermentation became crucial to avoid oxidation, which I learned through trial and error.

Sampling and testing became routine practices. Regularly checking the specific gravity helped me gauge fermentation progress, while pH levels ensured the environment remained conducive to yeast health. I discovered the importance of thorough documentation during this phase—keeping detailed records of each batch allowed me to analyze trends and refine my processes.

Addressing Common Challenges

Bulk fermentation brought its own set of challenges. Temperature fluctuations were a persistent issue. Large volumes often experienced temperature gradients, so I invested in effective cooling systems and insulating materials to maintain consistency. Managing sediment and trub was another hurdle. I opted for fermentation vessels with conical bottoms, which simplified the process of separating the mead from the sediment. Implementing racking techniques became an essential part of my process, ensuring that the final product remained clear and free from unwanted particles.

Contamination control was paramount. With larger volumes and more extensive equipment, the risk of contamination increased. I established rigorous cleaning protocols and used sanitizers to maintain a sterile environment. Regular inspections and maintenance of the fermentation vessels became routine practices to minimize any risk of contamination.

CHAPTER TEN

Post-Fermentation Management

As fermentation concluded, my focus shifted to post-fermentation management. Racking the mead into secondary vessels for aging was a crucial step. This not only helped in separating the mead from sediment but also allowed it to mature and develop its flavor profile. Monitoring the aging process, including temperature and storage conditions, became a key responsibility.

Clarification was another important phase. I experimented with various fining agents and filtration methods to achieve the desired clarity before bottling. Each step was aimed at ensuring that the final product was as polished and high-quality as possible.

Documentation and Record-Keeping

One of the most significant lessons I learned was the importance of documentation. Keeping detailed records of each fermentation batch, from ingredient quantities to fermentation conditions, helped me track performance and identify areas for improvement. Quality control logs became a vital resource for troubleshooting and ensuring consistency across batches.

Reviewing and analyzing production data regularly allowed me to optimize processes and make informed adjustments. Continuous improvement became a central theme, helping me refine my techniques and enhance the efficiency and quality of my bulk fermentation practices.

Managing fermentation in bulk has been an enlightening and rewarding journey. The transition from small-batch to large-scale production required me to embrace new technologies, refine my techniques, and continuously adapt to the challenges of larger volumes. Through careful planning, precise monitoring, and rigorous quality control, I've been able to scale up my mead-making process while maintaining the high standards I set for my product.

CHAPTER ELEVEN

TROUBLESHOOTING AND REFINING

Common Challenges in Mead Making

Fermentation Issues

A. Stuck Fermentation A stuck fermentation occurs when the yeast fails to complete the fermentation process, leaving residual sugars and an incomplete conversion of honey to alcohol. This issue can stem from various factors, including:

- **Nutrient Deficiency:** Yeast requires essential nutrients to thrive. A lack of nutrients, such as nitrogen or vitamins, can hinder yeast activity. To remedy this, use mead-specific yeast nutrient blends or additives like diammonium phosphate (DAP) to provide necessary nutrients.
- **Temperature Extremes:** Yeast activity is temperature-sensitive. Too high or too low temperatures can stress the yeast. Maintain fermentation temperatures within the optimal range for the yeast strain used, typically between 60-75°F (15-24°C), depending on the yeast type.

- **High Alcohol Content:** Excessive alcohol levels can inhibit yeast function. Ensure that the starting gravity of the must is within a reasonable range and use high-alcohol-tolerant yeast strains if aiming for higher alcohol content.

B. Off-Flavors Off-flavors in mead can result from various issues, including:

- **Sulfury Smells:** Sulfur compounds often arise from stressed yeast or insufficient aeration. Ensure proper aeration before pitching yeast and avoid over-stressing the yeast by monitoring fermentation conditions.
- **Vegetal or Yeasty Flavors:** These can be a sign of inadequate fermentation or improper aging. Allow sufficient time for fermentation to complete and age the mead properly to let off-flavors dissipate.

Clarification Issues

A. Cloudiness Cloudiness in mead, often due to suspended yeast or particulate matter, can be addressed through:

- **Racking:** Transfer the mead from one vessel to another, leaving sediment behind. This process, called racking, helps clarify the mead by removing sediment.
- **Fining Agents:** Use fining agents like bentonite or gelatin to clarify the mead. These agents bind with particles and settle them out, improving clarity.

B. Sediment Control Sediment can impact the appearance and flavor of mead. To minimize sediment issues:

- **Proper Handling:** Avoid excessive agitation of the mead and handle it gently during racking and bottling.
- **Use of Filter:** Employ a fine mesh filter or sterile filter during bottling to remove any remaining sediment.

Flavor Imbalances

A. Sweetness and Dryness The sweetness level of mead can be adjusted post-fermentation if needed:

- **Back-Sweetening:** If the mead is too dry, add additional honey or sweeteners to adjust the sweetness. Ensure that any added sugars are dissolved and thoroughly mixed to prevent uneven sweetness.
- **Blending:** Blend the mead with a sweeter batch or with honey syrup to achieve the desired sweetness level.

B. Overly Strong Flavors Sometimes, mead can develop strong flavors that overshadow the intended profile. Remedies include:

- **Dilution:** Dilute the mead with water or a neutral mead to balance strong flavors. This approach should be used cautiously to avoid overly diluting the mead.
- **Flavor Balancing:** Add complementary flavors, such as spices or fruits, to balance out dominant flavors and enhance the overall profile.

Carbonation Issues

A. Lack of Carbonation For meads that are intended to be carbonated, a lack of bubbles can be problematic. Address this issue by:

- **Priming Sugar:** Add priming sugar before bottling to encourage carbonation. This sugar will ferment in the bottle, producing CO_2.
- **Proper Sealing:** Ensure bottles are properly sealed to trap carbonation. Use bottles designed to withstand pressure if making sparkling mead.

B. Excessive Carbonation Over-carbonation can be controlled by:

- **Adjusting Priming Sugar:** Reduce the amount of priming sugar used

or adjust the fermentation time to prevent excessive CO2 production.
- **Monitoring Fermentation:** Keep an eye on fermentation activity to avoid over-fermentation and excessive carbonation.

Refining Mead for Optimal Quality

Aging and Maturation

Aging mead enhances its flavor complexity and smoothness. Allow mead to age in a cool, dark place for several months to a year, depending on the style. Taste periodically to monitor its development and ensure it reaches the desired flavor profile.

Flavoring and Enhancing

After primary fermentation, additional flavors can be introduced:

- **Fruits, Spices, and Herbs:** Add fruits, spices, or herbs to the mead during secondary fermentation or aging. This can create unique and complex flavor profiles.
- **Oak Aging:** Consider aging mead on oak chips or in oak barrels to impart additional flavors and complexity.

Balancing Acidity and Sweetness

Achieving the right balance between acidity and sweetness is key to a well-rounded mead. Use acid adjustments, such as citric acid or malic acid, to balance the overall flavor profile and ensure a harmonious taste experience.

Testing and Quality Control

Regularly test your mead for parameters like alcohol content, pH, and specific gravity. This helps ensure consistency and allows for adjustments before bottling.

Troubleshooting and refining mead making is an essential aspect of producing high-quality mead. By addressing common issues such as fermentation problems, clarification challenges, and flavor imbalances, mead makers can enhance their craft and achieve a superior product. Refinement techniques, including aging, flavoring, and balancing, further elevate the mead's quality and complexity.

Refining Techniques: Filtration, Fining, and Stabilization

Refining techniques are essential steps in mead making, focusing on enhancing clarity, improving flavor, and ensuring stability. Techniques such as filtration, fining, and stabilization address common issues related to haze, off-flavors, and fermentation completeness. Mastering these methods is crucial for producing a polished and high-quality mead.

1. Filtration

Filtration is a process used to remove unwanted particles, including yeast, sediment, and other particulates, from mead to achieve clarity and improve its overall appearance.

A. Types of Filtration

- **Gravity Filtration:** This method relies on gravity to pull mead through a filter medium. It is suitable for clearing larger particles and sediment. Gravity filtration is often used in the initial stages of clarification but may not be sufficient for fine filtration.
- **Pressure Filtration:** This technique uses pressure to force mead through a filter, effectively removing finer particles. Pressure filtration

can achieve a higher level of clarity and is commonly used in commercial mead making. It involves using a filter press or similar equipment to apply pressure.
- **Sterile Filtration:** Sterile filtration employs fine filters to remove yeast and microorganisms, preventing further fermentation or spoilage. It is typically used before bottling to ensure that no active yeast or contaminants are present. Sterile filters have very fine mesh sizes, often in the range of 0.45 to 0.65 microns.

B. Benefits of Filtration

- **Enhanced Clarity:** Filtration removes particulate matter that contributes to haze, resulting in a clearer and more visually appealing mead.
- **Improved Stability:** By removing yeast and microorganisms, filtration reduces the risk of unwanted fermentation or spoilage in the final product.
- **Consistent Quality:** Filtration helps ensure consistency in the appearance and quality of mead, making it more attractive to consumers.

2. Fining

Fining is a process that involves adding agents to mead to bind with and remove specific unwanted substances, such as proteins, polyphenols, or excess yeast, improving clarity and flavor.

A. Types of Fining Agents

- **Bentonite:** A clay-based fining agent effective at removing proteins and other particles that cause haze. Bentonite works by binding with proteins and forming a sediment that settles out of the mead. It is commonly used in both wine and mead making.
- **Gelatin:** A protein-based fining agent that helps clarify mead by binding

with particulate matter and yeast. Gelatin is typically used in small amounts and can be added during secondary fermentation or before bottling.
- **Isinglass:** Derived from fish bladders, isinglass is a traditional fining agent used to clarify mead by removing yeast and other fine particles. It is particularly effective for removing yeast haze but may not be suitable for vegan or vegetarian mead.
- **Pectin Enzymes:** Enzymes that break down pectin, a carbohydrate found in fruits that can cause haze. Pectinase is used during primary fermentation, especially in meads with high fruit content.

B. Benefits of Fining

- **Improved Clarity:** Fining agents help remove haze-causing particles, resulting in a clearer mead.
- **Enhanced Flavor:** By removing unwanted substances, fining agents can reduce off-flavors and improve the overall taste of mead.
- **Faster Aging:** Fining can expedite the aging process by clarifying mead more quickly, allowing it to be ready for consumption sooner.

3. Stabilization

Stabilization involves techniques used to prevent unwanted changes in mead after fermentation, such as further fermentation, spoilage, or flavor changes. Stabilization is crucial for ensuring the longevity and quality of the final product.

A. Methods of Stabilization

- **Sulfites:** Adding sulfites, such as potassium metabisulfite, helps inhibit the growth of unwanted microorganisms and prevent oxidation. Sulfites act as preservatives and can help stabilize mead by reducing the risk of

spoilage and off-flavors.
- **Sorbates:** Potassium sorbate is commonly used to prevent further fermentation by inhibiting yeast activity. It is effective at preventing refermentation in bottled mead, especially if back-sweetening has been done. Sorbates are typically used in conjunction with sulfites for optimal stabilization.
- **Cold Stabilization:** Chilling mead to near-freezing temperatures encourages the formation of tartrate crystals and other particles that can be removed through filtration. Cold stabilization helps ensure that the mead remains clear and stable over time.
- **Pasteurization:** Heat treatment can be used to kill yeast and microorganisms, preventing further fermentation. However, pasteurization can alter the flavor profile and is less commonly used in home mead making compared to sulfites and sorbates.

B. Benefits of Stabilization

- **Prevents Spoilage:** Stabilization techniques help prevent microbial contamination and spoilage, ensuring the mead remains safe and stable for consumption.
- **Maintains Flavor:** By preventing further fermentation and oxidation, stabilization helps preserve the intended flavor profile of the mead.
- **Extends Shelf Life:** Proper stabilization techniques can extend the shelf life of mead, allowing it to be stored and enjoyed over a longer period.

Experimentation and Iteration

In my mead-making journey, I've discovered that experimentation and iteration are pivotal for refining my craft. Each batch of mead is both a learning opportunity and a chance to innovate. My approach involves setting clear objectives, carefully designing experiments, and iterating based on results to continually enhance the quality of my meads.

When I start a new batch, I set specific goals. Whether it's perfecting a flavor profile, improving clarity, or boosting fermentation efficiency, having a clear aim helps me stay focused. For instance, I might experiment with different types of honey or yeast strains to see how they impact the final product. I document every detail meticulously—ingredient types, quantities, fermentation conditions, and any variations from my usual process. This record-keeping is crucial for tracking my progress and understanding what works.

As the batch progresses, I taste and analyze it regularly. Sensory evaluation allows me to gauge flavors and aromas, while technical analysis—like measuring gravity and alcohol content—provides insight into the fermentation process. If I notice that a certain honey adds an interesting flavor or that a clarification method yields a clearer mead, I make adjustments accordingly.

Iteration is where the real magic happens. Based on my findings, I refine my recipes and processes. I might tweak ingredient amounts, adjust fermentation conditions, or experiment with new clarification techniques. Starting with small batches helps me test these changes without committing too much at once. Once I find a combination that works, I scale up and apply the successful methods to larger batches.

Creativity is also a key part of my experimentation. I enjoy exploring unconventional flavors and techniques, whether it's infusing my mead with exotic spices or experimenting with different aging profiles. This creative aspect keeps the process exciting and allows me to develop unique meads that stand out.

Through continuous experimentation and careful iteration, I've learned to elevate my mead-making skills and produce exceptional meads. Each batch is a step toward refining my craft, and the journey of improvement is as rewarding as the final product.

IV

MEAD IN CULTURE AND COMMERCE

12

CHAPTER TWELVE

MEAD IN MODERN CULTURE

Historically, mead's origins are shrouded in the mists of time, with evidence suggesting it was consumed in various forms across ancient civilizations—from the Vikings in Scandinavia to the Greeks and Romans in the Mediterranean. It was revered not only for its intoxicating effects but also for its supposed health benefits and mystical qualities. Mead was often associated with rituals and celebrations, believed to be a gift from the gods or a potion to bestow good fortune.

In recent years, mead has re-emerged as a staple in the craft beverage world, paralleling the rise of artisanal beer, wine, and spirits. This revival is fueled by a blend of historical fascination and modern innovation. Contemporary mead-makers are reinventing the ancient beverage, experimenting with flavors, techniques, and ingredients that stretch the boundaries of traditional mead-making. This creative exploration reflects a broader trend within the craft beverage industry, where there is a continual push towards unique and personalized consumer experiences.

The modern mead renaissance is marked by a diversification of styles and flavors. Traditional meads, known for their simple sweetness, have given way to a spectrum of varieties including dry, semi-sweet, and sparkling meads. Mead-makers are now incorporating a range of fruits, spices, herbs, and even exotic ingredients, creating beverages that can rival, or even surpass, the complexity of wines and beers. This innovation has brought mead into the spotlight, appealing to adventurous drinkers seeking new tastes and experiences.

Moreover, mead's resurgence is closely linked with the growing interest in sustainable and locally sourced products. Many modern meaderies pride themselves on using high-quality, local honey, which not only supports regional beekeepers but also ensures that each batch of mead has a unique flavor profile tied to its geographical origin. This emphasis on local ingredients resonates with contemporary consumers who are increasingly conscientious about the provenance and environmental impact of their food and drink choices.

In the context of social media and digital culture, mead has found a new audience among enthusiasts and hobbyists who share their mead-making experiences, recipes, and discoveries online. Platforms like Instagram, YouTube, and specialized forums have become virtual mead halls, where knowledge is exchanged and new trends are born. This digital presence has contributed to the democratization of mead-making, allowing enthusiasts to connect with experts and fellow hobbyists, thereby fostering a vibrant and supportive community.

Despite its newfound popularity, mead still contends with a degree of obscurity compared to more mainstream alcoholic beverages. This can partly be attributed to historical factors, such as the dominance of beer and wine in the Western world and mead's association with niche or historical contexts. However, this is changing as more people discover mead through craft beverage festivals, local meaderies, and educational resources that highlight its rich heritage and contemporary applications.

In summary, mead's place in modern culture is a testament to the enduring appeal of historical beverages reimagined for contemporary tastes. Its revival is a product of both historical reverence and modern innovation, reflecting a broader trend towards artisanal and unique alcoholic experiences.

13

CHAPTER THIRTEEN

MEAD AND LEGAL CONSIDERATIONS

Historically, mead has been brewed and consumed for thousands of years. Ancient civilizations such as the Greeks, Romans, and Vikings held mead in high esteem, associating it with celebration, fertility, and divine favor. In modern times, mead has experienced a resurgence in popularity, driven by a growing interest in artisanal and craft beverages. This revival has brought mead into the mainstream but has also introduced new legal considerations.

Licensing and Regulation

One of the primary legal considerations for mead production is obtaining the appropriate licenses. In many countries, including the United States, mead is classified as an alcoholic beverage, and its production is subject to specific regulatory frameworks.

United States

In the U.S., mead falls under the jurisdiction of the Alcohol and Tobacco Tax and Trade Bureau (TTB). Mead producers must obtain federal permits for the production, distribution, and sale of mead. This includes applying for a Federal Basic Permit and adhering to regulations concerning labeling, advertising, and record-keeping. Additionally, each state has its own regulations regarding the sale and distribution of alcoholic beverages. Some states have laws that are particularly stringent, affecting everything from the production volume to the sale of mead.

European Union

In the European Union, mead production is regulated under the broader category of alcoholic beverages. Regulations vary by country but generally include compliance with health and safety standards, labeling requirements, and taxation laws. For example, in countries like the United Kingdom and Germany, mead producers must adhere to specific standards regarding ingredients and production processes to ensure consumer safety and product quality.

Other Regions

In countries outside these regions, the legal landscape can be quite different. In some places, mead may be less regulated or classified under different categories, such as specialty or craft beverages. Producers must research local laws to ensure compliance with all relevant regulations.

Taxation

Taxation is another critical area of legal consideration. Alcoholic beverages, including mead, are subject to excise taxes imposed by governments. These taxes are often based on the volume and alcohol content of the beverage. For commercial producers, calculating and remitting these taxes accurately is crucial to avoid legal issues and financial penalties.

In the U.S., for example, the TTB sets tax rates for mead, which are subject to change based on federal legislation. Producers must keep detailed records of production and sales to ensure accurate tax reporting. In the EU, taxation rates vary by country, and producers must comply with both national and EU-wide tax regulations.

Homebrewing Regulations

For homebrewers, the legal considerations are somewhat different but still significant. In many jurisdictions, homebrewing for personal use is permitted within certain limits. However, selling homebrewed mead without proper licensing is generally illegal. Homebrewers must be aware of the quantity limits imposed by local laws and any restrictions on the distribution or sale of homebrewed beverages.

United States

In the U.S., federal law permits homebrewing up to 100 gallons per year for individuals and 200 gallons per year for households with two or more adults. However, laws can vary by state, and some states have additional regulations or restrictions. Homebrewers should familiarize themselves with state and local laws to ensure they are compliant.

European Union

In the EU, regulations on homebrewing vary by country. Some countries have more relaxed rules, while others impose stricter limits on the quantity of homebrewed beverages. Homebrewers must also be cautious about sharing or selling their products, as this often requires a license or adherence to specific regulations.

Labeling and Advertising

CHAPTER THIRTEEN

Labeling and advertising are essential components of mead production that involve legal considerations. Accurate labeling ensures that consumers are informed about the product they are purchasing, including its alcohol content, ingredients, and origin. Misleading or incorrect labeling can result in legal penalties and damage to the producer's reputation.

United States

In the U.S., the TTB regulates labeling requirements for alcoholic beverages. Labels must include information such as the brand name, type of beverage, alcohol content, and the name and address of the producer. Additionally, labels must meet specific guidelines regarding ingredient statements and health warnings.

European Union

In the EU, labeling regulations are similarly stringent. Labels must provide detailed information about the product, including its ingredients, allergens, and alcohol content. Additionally, some countries require specific language or certification marks on labels.

Understanding Local Laws: Homebrewing vs. Commercial Production

Homebrewing Laws

Homebrewing refers to the practice of making alcoholic beverages, including mead, for personal use. Laws governing homebrewing are generally more lenient compared to those for commercial production, but they still impose certain restrictions.

Legal Limits

Quantity Restrictions

In the US, Under federal law, individuals can homebrew up to 100 gallons of beer or mead per year per adult (up to 200 gallons for households with two or more adults). For instance, in a state like California, this limit is generally respected, and homebrewers can legally produce mead within these quantities without a commercial license.

For example, A homebrewer in California who produces 50 gallons of mead for personal use is within the legal limit. However, if the same individual wishes to produce more than 100 gallons, they would need to obtain the necessary permits and licenses for commercial production.

Non-Commercial Use

Restrictions on Sales

Homebrewed mead cannot be sold or distributed commercially. Any exchange of homebrew, whether for money or barter, is typically prohibited. For example, a homebrewer in New York cannot sell their mead at a local farmers' market, as this would require a commercial license.

For example, A homebrewer in Texas who wants to share their mead with friends at a gathering can do so without issue. However, if they try to sell bottles at a local event, they would be violating local regulations.

Regulatory Compliance

Local Regulations

Homebrewing laws can vary by state and municipality. Some local jurisdictions may have additional restrictions or requirements for homebrewers. For example, in some counties in Florida, homebrewing may require notification to local authorities to ensure compliance with health and safety standards.

Permits and Notifications

Basic Permits

In most places, homebrewers do not require permits as long as they adhere to quantity limits and do not sell their product. However, it's advisable to check with local authorities to confirm specific requirements.

For example, In Oregon, homebrewers generally do not need a permit as long as they stay within the federal limits and do not engage in commercial activities. However, local zoning laws may affect where homebrewing can occur, particularly if it involves substantial equipment or storage.

Commercial Production Laws

Commercial mead production involves producing mead for sale, which entails a more stringent regulatory framework compared to homebrewing. This framework includes federal, state, and local regulations that govern licensing, taxation, and operational standards.

Licensing and Permits

Federal Licensing

Alcohol and Tobacco Tax and Trade Bureau (TTB): Commercial mead producers must obtain a Federal Basic Permit from the TTB. This permit allows them to legally produce and sell mead. The application process involves detailed background checks, financial disclosures, and adherence to federal regulations.

State Licensing

State Liquor Control Boards

In addition to federal permits, commercial mead producers must obtain state-level licenses. Each state has its own regulatory body that oversees alcohol production and distribution. Licensing requirements vary by state and may include additional permits for distribution and retail sales.

For example, A meadery in Colorado must obtain a state liquor license from the Colorado Liquor Enforcement Division. This involves meeting state-specific requirements for production facilities, labeling, and record-keeping.

Local Permits

Zoning and Health Permits

B Local regulations may require zoning permits and health inspections to ensure that the production facility meets local codes. This can include compliance with building codes, health and safety standards, and environmental regulations.

Example: A meadery in San Francisco would need to secure a zoning permit to operate in a designated industrial area and pass a health inspection to ensure sanitary conditions in their production facility.

Taxation and Reporting

Federal Excise Taxes

Tax Rates

Commercial mead producers must pay federal excise taxes based on the alcohol content and volume of their mead. Accurate reporting and timely payment are essential for compliance.

Example: A meadery producing 1,000 gallons of mead with an alcohol content of 8% would calculate federal excise taxes based on the TTB's current tax rates. They must file regular tax returns and remit payment to the TTB.

State Taxes

State Excise and Sales Taxes: In addition to federal taxes, states impose their own excise and sales taxes on alcoholic beverages. Producers must comply with these tax laws and file appropriate returns.

Example: A meadery in Washington State must pay both state excise tax and sales tax on their mead. They need to account for these taxes in their pricing and financial reporting.

Record-Keeping

Detailed Records: Commercial producers are required to maintain detailed records of production, inventory, and sales. These records must be accurate and accessible for inspection by regulatory agencies.

Example: A meadery in New York must keep detailed logs of each batch of mead produced, including ingredients, production dates, and sales transactions, to ensure compliance with both federal and state regulations.

Key Differences Between Homebrewing and Commercial Production

Regulatory Scope

Homebrewing

Generally regulated by federal limits on quantity and prohibition on sales. Local laws may impose additional restrictions but are less comprehensive.

Commercial Production

Subject to extensive federal, state, and local regulations, including licensing, taxation, and health standards.

Permitting

Homebrewing

Minimal to no permits required for personal use, with focus on quantity limits and non-commercial distribution.

Commercial Production

Requires multiple permits and licenses from federal, state, and local authorities, including detailed applications and inspections.

Taxation

Homebrewing

No excise taxes or sales taxes for personal use.

Commercial Production

Subject to federal excise taxes, state excise taxes, and potentially sales taxes, with detailed reporting requirements.

Understanding the legal distinctions between homebrewing and commercial mead production is crucial for compliance and successful operation. Homebrewing laws focus on personal use and limit quantities, while commercial production involves a comprehensive regulatory framework, including multiple licenses, rigorous taxation, and detailed record-keeping.

CHAPTER THIRTEEN

Licensing and Regulations for Commercial Mead Makers

Commercial mead production involves navigating a complex web of licensing and regulatory requirements. These regulations are essential to ensure consumer safety, product quality, and legal compliance.

Federal Licensing and Regulations

Alcohol and Tobacco Tax and Trade Bureau (TTB)

In the United States, the Alcohol and Tobacco Tax and Trade Bureau (TTB) is the primary federal agency overseeing the production, distribution, and sale of alcoholic beverages, including mead. Commercial mead makers must comply with TTB regulations, which encompass several key areas:

Federal Basic Permit

- **Application**: To legally produce and sell mead, a business must obtain a Federal Basic Permit from the TTB. This involves submitting an application that includes details about the business, its location, and its production processes. The application process includes background checks and financial disclosures.
- **Approval**: The TTB reviews the application to ensure the business meets all legal requirements, including compliance with federal alcohol laws and regulations. Approval can take several months, depending on the complexity of the application and the TTB's workload.

Formula Approval

- **Required for New Recipes**: If a mead maker uses a new or unique recipe that includes ingredients not typically used in mead production, they must seek formula approval from the TTB. This ensures that the

product meets federal standards for alcoholic beverages.
- **Submission**: The process involves submitting detailed information about the ingredients and production methods. The TTB reviews this information to ensure the formula complies with federal regulations.

Label Approval

- **Labeling Requirements**: The TTB regulates the labeling of mead to ensure it meets federal standards. Labels must include specific information such as the brand name, type of beverage, alcohol content, and the name and address of the producer.
- **Submission and Approval**: Mead makers must submit label samples to the TTB for approval before they can be used on bottles. The TTB reviews labels to ensure they are accurate and comply with federal labeling requirements.

Record-Keeping and Reporting

- **Detailed Records**: Commercial mead makers are required to maintain detailed records of production, inventory, and sales. This includes tracking the volume of mead produced, ingredients used, and sales transactions.
- **Reports and Returns**: Producers must file periodic reports with the TTB, including excise tax returns and production reports. Accurate and timely reporting is essential for compliance and avoiding penalties.

Alcohol Excise Tax

Tax Rates

- **Excise Tax Rates**: The TTB sets federal excise tax rates for alcoholic beverages, including mead. Tax rates can vary based on the alcohol

content and volume of the product.
- **Tax Calculation**: Mead makers must calculate and remit excise taxes based on these rates. Understanding and applying the correct tax rates is crucial for compliance.

Tax Payments

- **Frequency**: Excise taxes are typically paid on a monthly or quarterly basis. Producers must adhere to the payment schedule established by the TTB.
- **Accuracy**: Accurate tax reporting and payment are essential to avoid fines and legal issues.

State Licensing and Regulations

Each state has its own regulatory framework governing the production and sale of alcoholic beverages. State regulations can vary significantly, so mead makers must be familiar with the laws in the states where they operate.

State Liquor Control Board or Commission

- **State Permits**: In addition to federal licensing, mead makers must obtain state permits or licenses to operate legally. This may include separate licenses for production, distribution, and retail sales.
- **Application Process**: The application process for state permits often involves submitting detailed information about the business, its facilities, and its production processes. Some states also require background checks and financial disclosures.

State Taxation

- **Excise Taxes**: States impose their own excise taxes on alcoholic beverages, which can differ from federal rates. Mead makers must comply with state tax laws and remit taxes according to state regulations.
- **Sales Tax**: In addition to excise taxes, mead makers may be required to collect and remit sales tax on retail sales, depending on state laws.

Distribution and Retail Regulations

- **Distribution Licenses**: States may have specific requirements for distributing alcoholic beverages. This can include obtaining distribution licenses and adhering to state-specific distribution laws.
- **Retail Sales**: If mead makers sell directly to consumers, they must comply with state regulations governing retail sales, including licensing and reporting requirements.

Local Regulations

Local jurisdictions may impose additional regulations on mead production and sales. These regulations can vary widely and may include zoning laws, health and safety requirements, and local licensing.

Zoning and Land Use

- **Zoning Laws**: Local zoning laws may dictate where mead production facilities can be located. Producers must ensure their facilities comply with local zoning regulations and obtain any necessary permits.
- **Land Use Permits**: Some areas may require land use permits or approvals for operating a commercial mead production facility.

Health and Safety

- **Health Inspections**: Local health departments may conduct inspections of production facilities to ensure compliance with health and

safety standards.
- **Sanitation Requirements**: Mead makers must adhere to local sanitation requirements to ensure product safety and prevent contamination.

Navigating Taxation and Distribution

Taxation is a significant aspect of mead production that impacts profitability and regulatory compliance. The taxation landscape for mead can be complex, involving federal, state, and sometimes local tax laws.

Federal Taxation

In the United States, the Alcohol and Tobacco Tax and Trade Bureau (TTB) is responsible for federal excise taxes on alcoholic beverages, including mead. The tax rate depends on the type and alcohol content of the beverage. Mead is taxed similarly to other alcoholic beverages but with some variations depending on its classification.

1. **Excise Tax Rates**: As of the latest regulations, mead is taxed at a rate based on its alcohol content. The TTB applies different rates for mead with varying alcohol levels. For example, a lower tax rate may apply to mead with a lower alcohol content compared to stronger meads.
2. **Filing and Payment**: Producers must file tax returns and pay excise taxes on a periodic basis, typically monthly or quarterly. The TTB requires detailed records of production, sales, and inventory to ensure accurate tax reporting.
3. **Credits and Reductions**: Some producers may be eligible for tax credits or reductions based on specific circumstances, such as producing mead in smaller quantities or operating under certain conditions. It's important to stay informed about any available credits and apply them

as applicable.

State and Local Taxation

In addition to federal taxes, mead producers must comply with state and local tax regulations. These taxes can vary widely depending on the location.

1. **State Excise Taxes**: Each state sets its own excise tax rates on alcoholic beverages. These rates can differ based on the type of beverage and its alcohol content. Producers must research and adhere to the tax rates in each state where they operate or distribute.
2. **Local Taxes**: Some local jurisdictions impose additional taxes on alcoholic beverages. Producers should be aware of any local tax obligations and ensure compliance to avoid penalties.
3. **Reporting Requirements**: States and localities may have specific reporting requirements, including frequency and format of tax filings. Producers should maintain accurate records and follow local guidelines to meet these requirements.

Distribution

Distribution involves the logistics of getting mead from the production facility to consumers, including wholesalers, retailers, and directly to customers. Effective distribution management is crucial for maximizing reach and profitability.

Distribution Channels

1. **Wholesalers**: Wholesalers purchase mead in bulk and distribute it to retailers or other businesses. Establishing relationships with wholesalers can expand market reach and streamline distribution. Producers should research potential wholesalers, negotiate terms, and ensure compliance with any contractual obligations.

2. **Retailers**: Retail distribution involves selling mead to retail outlets, such as liquor stores, supermarkets, or specialty shops. Producers must navigate retailer agreements, including pricing, placement, and promotional support. Building strong relationships with retailers can enhance visibility and sales.
3. **Direct-to-Consumer**: Some producers choose to sell mead directly to consumers through online sales or on-site tasting rooms. This approach allows for greater control over the sales process and customer interactions. However, direct-to-consumer sales are subject to various regulations, including shipping laws and sales tax.

Distribution Logistics

1. **Shipping and Handling**: Effective logistics are essential for ensuring that mead is delivered in good condition and within legal parameters. Producers must comply with shipping regulations, including packaging, labeling, and handling requirements. Additionally, they must manage shipping costs and delivery timelines to meet customer expectations.
2. **Compliance with Laws**: Distribution is subject to numerous regulations, including state and local alcohol laws. These laws can affect shipping routes, sales limits, and licensing requirements. Producers must stay informed about applicable regulations and ensure compliance throughout the distribution process.
3. **Inventory Management**: Managing inventory effectively is crucial for balancing supply and demand. Producers should use inventory management systems to track production, sales, and distribution. This helps prevent stockouts, overproduction, and financial losses.

Marketing and Sales

1. **Marketing Strategies**: Effective marketing strategies can enhance distribution efforts by increasing brand awareness and attracting

customers. Producers should develop marketing campaigns that resonate with target audiences and promote their mead through various channels, including social media, events, and collaborations.
2. **Sales Agreements**: Sales agreements with wholesalers, retailers, and distributors should be carefully negotiated and documented. These agreements outline terms related to pricing, delivery schedules, and promotional activities. Clear and mutually agreed-upon terms help prevent disputes and ensure smooth operations.

14

CHAPTER FOURTEEN

STARTING A MEADERY

Starting a meadery is a journey that combines the love of an ancient craft with the realities of running a business. It's about turning a deep appreciation for mead, one of the oldest alcoholic drinks in human history, into something that other people can enjoy. But it's not just about the drink. It's about creating a space—physical or metaphorical—where tradition, creativity, and entrepreneurship intersect.

If you're thinking of opening a meadery, you're probably someone who appreciates mead not just as a beverage but as a piece of cultural heritage. Mead has been around for thousands of years, and its resurgence in recent years is exciting for people looking to introduce others to its unique and diverse flavors. However, making mead on a commercial scale requires much more than just a passion for the drink.

At the core of any meadery is the mead itself, and that means understanding the craft intimately. Mead starts with three basic ingredients: honey, water, and yeast. While these may sound simple, the choices you make around each will deeply affect the final product.

Planning and Strategies

Starting a meadery requires more than just a passion for mead; it's a business endeavor that needs careful planning and thoughtful strategies. From understanding the mead market to sourcing high-quality ingredients, every step needs to be methodically approached to ensure your business's success. Below are the key planning elements and strategies to help guide you through the process, along with examples that will make these concepts more tangible.

1. Developing a Solid Business Plan

Before you can turn your mead-making hobby into a commercial venture, you need a comprehensive business plan. This document will serve as the blueprint for your meadery and should include details such as startup costs, revenue projections, target market, and marketing strategies.

For example, let's say you plan to focus on high-end, premium meads made from locally sourced honey and organic ingredients. Your business plan should outline:

- **Costs**: How much will it cost to purchase honey, yeast, water, and spices? How much will equipment like fermentation tanks, bottling machines, and a labeling system set you back?
- **Revenue Projections**: Based on the price per bottle and estimated sales volume, how much revenue can you expect in the first year? How long will it take to break even?
- **Market Research**: Who is your target audience? Are they local craft beverage enthusiasts, or are you aiming to sell nationwide? Understanding who your customers are will help shape your product offerings and marketing strategies.
- **Marketing Strategy**: How will you reach your audience? Will you start with local farmer's markets, or do you plan to open a tasting room?

CHAPTER FOURTEEN

What role will social media and online sales play?

By thinking through these elements ahead of time, you'll have a clearer picture of how your meadery will operate and what steps are necessary to grow it over time.

2. Securing Capital and Funding

Starting a meadery can be costly, and many entrepreneurs seek outside funding to cover their startup costs. Some opt for traditional bank loans, while others might turn to investors or crowdfunding platforms like Kickstarter to raise money.

Let's look at an example of crowdfunding. Imagine you want to raise $50,000 to cover equipment and initial production costs. You could launch a Kickstarter campaign where you offer backers rewards in exchange for their support. For instance, at the $50 level, backers could receive a limited-edition bottle of your first mead, while at the $500 level, they could get a behind-the-scenes tour of your meadery. Crowdfunding not only raises funds but also helps build an initial customer base.

3. Choosing the Right Location

Location is everything, and this decision will depend heavily on your business model. If you plan to focus on local sales and open a tasting room, you'll want to choose a location with good foot traffic and proximity to local attractions or popular tourist areas.

For example, if you set up your meadery near a wine or brewery trail, you could attract craft beverage enthusiasts who are already visiting the area. Or, if you're targeting a more urban demographic, setting up in a trendy part of town known for its craft breweries and artisan food vendors could help drive sales through walk-in traffic.

If your primary focus is on wholesale or online sales, then the location may be less about foot traffic and more about having enough space for production and storage at a reasonable cost. In this case, you might consider leasing a warehouse or setting up in a light industrial zone where rent is more affordable.

4. Sourcing Ingredients

One of the most important decisions you'll make is where to source your honey. Honey is the soul of mead, and the quality and variety you choose will greatly impact the flavor of your mead. Establishing relationships with local beekeepers can ensure a steady, high-quality supply, but it also gives you a unique selling point—local, artisanal ingredients that customers love to support.

For example, if your meadery is located in a region known for its wildflower honey, you could craft meads that celebrate the local terroir, just like wineries do with grapes. You could even name your meads after local landmarks or flora to reinforce the connection to your area, adding both authenticity and a story behind each bottle.

Beyond honey, if you plan to make flavored meads—known as melomels or metheglins—sourcing fresh, seasonal fruits and spices is key. Partnering with local farmers or organic suppliers can elevate the quality of your product and create exciting, limited-edition releases that appeal to a wide audience.

5. Navigating Licensing and Legal Requirements

One of the biggest hurdles in starting a meadery is navigating the various licensing requirements. Alcohol production is heavily regulated, and you'll need to secure federal, state, and local permits before you can start selling mead. In the U.S., for example, you'll need approval from the Alcohol and Tobacco Tax and Trade Bureau (TTB), and your state may have additional licensing requirements as well.

Consider a scenario where you've planned everything, but the licensing process takes longer than expected. This is where your planning comes in handy. You'll want to build a buffer into your timeline and have contingency plans in place, such as securing pre-orders or hosting mead-making workshops to keep interest alive while waiting for approvals.

6. Setting Pricing and Distribution Strategy

Pricing your mead is a critical decision that affects your profitability, brand perception, and market reach. You'll need to strike a balance between covering your production costs and offering a competitive price that reflects the quality of your product. Premium meads, especially those made from organic honey or infused with rare ingredients, can command higher prices, but they must be positioned and marketed correctly.

Let's consider an example: Suppose you've developed a line of high-end meads with a $25 price point per bottle, which is higher than your local competitors. In this case, your distribution strategy might focus on exclusive channels, such as high-end wine shops, gourmet food stores, and online platforms targeting affluent customers. You may also focus on building relationships with restaurants and bars that specialize in craft beverages.

Alternatively, if you're producing more approachable, everyday meads with a $15 price point, you might aim for broader distribution. This could include local liquor stores, farmer's markets, and craft beer festivals, where the price point is in line with the average consumer's budget.

7. Marketing and Building a Brand

Your brand will be what sets your meadery apart from others. This is more than just a logo or label—it's about the story behind your meadery, your values, and the experience you provide to your customers. Are you a small, family-run meadery focusing on traditional methods? Or are you a cutting-edge producer experimenting with unusual flavors and techniques?

For example, if you decide to focus on sustainability, you can market your meadery as eco-conscious, using organic ingredients and sustainable production methods. You might even package your meads in recyclable or eco-friendly bottles, appealing to environmentally aware consumers. Another meadery might emphasize their use of wild fermentation techniques, creating a more experimental brand that attracts craft beverage enthusiasts looking for something new and bold.

Marketing strategies could include hosting tasting events, offering tours of your meadery, collaborating with local artisans or breweries, and building a strong presence on social media platforms like Instagram and Facebook.

Scaling Up: Managing Growth and Expansion

Managing the growth and expansion of a meadery is an exciting, yet complex stage in your business journey. It's that moment when you realize your small operation is gaining traction, and the demand for your product is steadily increasing. However, with growth comes the need for careful management—expanding too quickly or without proper foresight can lead to challenges that might hinder your success.

When you first start your meadery, it might feel manageable. You have a small team (or maybe it's just you), a clear production process, and enough equipment to meet the current demand. But as word spreads, and you begin to build a loyal customer base, the orders start to pile up. You find yourself running low on mead, and your production schedule feels tighter with every batch. This is where expansion becomes not just an option, but a necessity. Yet, it's not as simple as doubling your production or hiring more hands. Growth requires a strategic approach to ensure you can scale sustainably and maintain the quality and uniqueness of your product.

Understanding the Signs of Growth

The first step in managing expansion is recognizing when it's the right time. Growth isn't just about increased sales—it's also about ensuring that the demand for your product is consistent enough to justify scaling up.

Imagine you've been selling out of your mead at local farmer's markets every weekend, and the same customers keep coming back, asking for more. Your tasting room is busy, with people not only buying bottles to take home but also recommending your meads to friends. You've even had a few inquiries from local restaurants or stores looking to carry your product. These are all positive signs of growth. It's not just a one-time spike in interest, but a pattern of increasing demand that shows your meadery has a solid market presence.

However, it's important to distinguish between short-term success and sustainable growth. Perhaps you've had a particularly successful launch or seasonal event that boosted sales, but will that momentum carry forward throughout the year? Expanding your meadery based on a temporary surge in demand can lead to overextending resources, which could strain your operations when things return to normal.

Strategizing for Scalable Production

Once you've determined that your growth is consistent, the next step is to scale your production. This requires both investment and planning. Simply making larger batches of mead isn't always as straightforward as it sounds—each element of production, from sourcing more honey to managing fermentation times and bottling, will need to be adapted.

For instance, let's say your meadery is currently producing 500 bottles per month, but the demand is pushing you to double that. You can't just assume doubling your ingredients will result in the same quality. Scaling up fermentation processes, particularly with yeast, can sometimes change the character of your mead. Larger fermentation tanks might behave differently from the smaller ones you've been using, and you'll need to experiment with larger batches to ensure that the taste, aroma, and quality remain consistent.

Additionally, as you expand production, sourcing enough honey can become a challenge. In the early days, you may have been able to work with a local beekeeper or two, but as you grow, you'll need more suppliers to keep up with your demand. It's crucial to maintain the same quality of ingredients as you scale—after all, your customers are returning for that specific flavor and quality that your mead offers. Securing reliable and scalable honey sources is a key part of managing your growth.

Investing in Equipment and Space

Scaling up also means investing in new equipment and possibly expanding your physical space. Larger fermenters, bottling machines, and storage tanks are necessary to accommodate increased production. This requires significant capital, so managing your cash flow becomes essential. It's easy to get carried away with expansion and over-invest in equipment before your business is truly ready to support it.

Take, for example, a meadery that expands too quickly by purchasing multiple new fermenters and bottling lines. If the demand doesn't continue to grow as expected, those expensive pieces of equipment could sit idle, tying up valuable capital that could have been used elsewhere in the business. A more cautious approach would be to incrementally increase your equipment as sales steadily grow, ensuring that each new purchase is backed by real demand.

Expansion might also mean moving to a bigger facility. This can be a significant expense, so it's important to weigh the costs against the benefits. Could you modify your current space to maximize production before making a move? Or is a new facility crucial to meeting your long-term goals?

Building a Team for Growth

As your meadery grows, you'll inevitably need to bring on more staff to help with production, marketing, sales, and customer service. Managing a team is a different skill set from managing production, and it requires thoughtful hiring and clear communication. The people you bring on board will shape the future of your business, so finding individuals who share your vision and passion for mead is essential.

Start by identifying which areas of your meadery need the most support. For instance, if production is becoming overwhelming, you may need an assistant brewer to help with fermentation and bottling. If marketing and sales are where you need help, hiring someone with experience in the craft beverage industry can open new opportunities for distribution and brand recognition.

However, bringing in new team members doesn't just mean handing off tasks—it also means creating a company culture that supports growth. The last thing you want is to lose the personal, handcrafted essence of your meadery in the rush to expand. Take time to train new staff on your production methods, quality standards, and the story behind your brand. This ensures that everyone is on the same page and working towards the same goals, even as the company scales.

Expanding Distribution Channels

As your production capacity increases, it's only natural to explore new distribution channels. While you may have started selling locally, growth often means expanding into regional or even national markets. This could involve partnering with distributors, selling your mead in liquor stores or grocery chains, or expanding your online sales.

However, expanding distribution brings its own challenges. For example, working with distributors means less control over how your product is marketed and sold. You'll also need to consider shipping logistics and how to maintain product quality over longer distances. When expanding to new markets, it's critical to ensure that the story behind your mead and the quality of your product remain consistent, even if it's no longer being sold directly by you.

You may also want to consider diversifying your product line. Offering seasonal or limited-edition meads can help generate buzz and bring in new customers. However, it's important not to stretch yourself too thin by introducing too many new products at once. Focus on perfecting a few core offerings before expanding your range.

Maintaining Quality During Expansion

Perhaps the most important part of managing growth is maintaining the quality and authenticity that attracted customers in the first place. As you scale, there's a natural temptation to cut corners or find ways to make the production process more efficient, but at the cost of the product. In the craft beverage industry, your reputation is everything, and customers will quickly notice if your mead starts to lose the charm that made it special.

For example, if you've built your brand on using organic, locally sourced honey, sticking to those values as you grow can be challenging. But compromising on quality for the sake of efficiency can erode customer trust and loyalty. You'll need to find suppliers and partners who align with your mission, even as your demands increase.

Managing growth and expansion in a meadery is a delicate balancing act. It requires thoughtful planning, smart investments, and a commitment to maintaining quality as you scale. By recognizing the signs of sustainable growth, carefully scaling production, building a supportive team, and expanding distribution strategically, you can grow your meadery without losing the essence of what made it successful in the first place.

V

MEAD RECIPES

15

CHAPTER FIFTEEN

TRADITIONAL MEAD RECIPES

1. Traditional Viking Mead (Honey Wine)

Origin: Scandinavia, 8th-11th centuries

Ingredients:

- Honey
- Water
- Yeast (wild or cultivated)
- Optional: Spices (e.g., cinnamon, cloves), herbs (e.g., sage, thyme)

Method: To create a Viking mead, I begin by mixing honey with water. I use a ratio of about 1 part honey to 4 parts water, adjusting based on desired sweetness and strength. This mixture, known as the must, is then transferred into a fermentation vessel, typically a wooden barrel. I add yeast to the must; historically, this might have been wild yeast from the environment, though I can also use a cultivated yeast for more controlled fermentation.

Next, I cover the vessel and allow the mixture to ferment. The fermentation process usually takes several weeks to months, depending on temperature and yeast activity. I check the mead periodically for signs of fermentation, such as bubbling or a change in aroma. Once fermentation slows or stops, indicating that the sugars have been converted to alcohol, I transfer the mead to another container for aging. The aging process can also last several months, allowing the flavors to mature and develop complexity.

Significance: Viking mead was integral to Norse culture, consumed during feasts and ceremonial gatherings. Its simple ingredients and method reflect the Vikings' resourcefulness and their deep appreciation for mead as a symbol of celebration and hospitality.

2. Traditional Ethiopian Tej

Origin: Ethiopia, ancient times

Ingredients:

- Honey
- Water
- Tej (or gesho) leaves (a local variety of hops)
- Optional: Spices like ginger or cloves

Method: To make Tej, I first combine honey with water in a large container, typically a clay vessel. I use about 1 part honey to 3 parts water. I then add gesho leaves to the mixture; these leaves act as a natural bittering agent and are crucial for the authentic flavor. For additional complexity, I might

CHAPTER FIFTEEN

include spices like ginger or cloves.

After mixing the ingredients, I cover the vessel with a cloth or lid and allow the mixture to ferment. The fermentation process usually takes several weeks, though it can extend to several months. During fermentation, I monitor the vessel for bubbling or changes in the aroma. Once the fermentation has slowed or stopped, I strain out the gesho leaves and spices, then transfer the Tej into bottles or jars for further aging.

Significance: Tej is a key element of Ethiopian cultural and social life, prominently featured in celebrations and rituals. The use of gesho leaves highlights local agricultural practices and flavor preferences, while the slow fermentation process allows for a richly developed taste.

3. Traditional Chinese Mijiu

Origin: China, ancient times

Ingredients:

- Honey
- Water
- Rice or glutinous rice
- Yeast (specific strain used in traditional Chinese brewing)

Method: For Mijiu, I start by preparing the rice. I steam it until it is fully cooked and sticky. In a separate container, I mix honey with water in a ratio of about 1 part honey to 4 parts water. Once the rice has cooled slightly, I add it to the honey-water mixture.

Next, I introduce a special yeast known as "qu" into the mixture. This yeast is crucial for fermentation and imparts unique flavors to the Mijiu. I stir the mixture thoroughly to ensure the yeast is well distributed. I then transfer the mixture into a ceramic jar or crock and cover it to prevent contamination.

Fermentation takes place over several weeks. During this period, I periodically check the jar for signs of fermentation, such as bubbles or changes in smell. Once fermentation appears to be complete, I strain out the rice and transfer the Mijiu into bottles for aging.

Significance: Mijiu is deeply rooted in Chinese brewing traditions, reflecting the historical importance of rice cultivation and fermentation techniques in Chinese culture. The use of specialized yeast and rice highlights the sophisticated methods developed to produce this distinctive beverage.

4. Traditional Welsh Metheglin

Origin: Wales, medieval times

Ingredients:

- Honey
- Water
- Spices (e.g., cinnamon, cloves, ginger)
- Herbs (e.g., mint, thyme)

Method: To prepare Metheglin, I start by mixing honey with water, typically in a 1:4 ratio. I then bring the mixture to a boil in a large pot, adding a selection of spices and herbs. Common choices include cinnamon, cloves, ginger, mint, and thyme. Boiling the mixture extracts the flavors from the spices and herbs, enriching the mead.

CHAPTER FIFTEEN

Once the mixture has been simmered and the spices have infused their flavors, I let it cool to room temperature. After cooling, I transfer the liquid into a fermentation vessel and add yeast. The vessel is then covered and left in a cool, dark place to ferment. This process usually takes several weeks. I monitor the fermentation for activity and transfer the mead to a secondary container for aging once fermentation has slowed.

Significance: Metheglin reflects Welsh traditions and the use of spices and herbs in brewing. It was enjoyed during festivals and special occasions, embodying local customs and taste preferences. The rich flavor profile of Metheglin showcases the importance of spice blending in Welsh mead-making.

Basic Traditional Mead

Basic traditional mead is a straightforward and ancient alcoholic beverage made primarily from honey, water, and yeast.

Ingredients

1. **Honey**: The primary ingredient in mead, honey is the source of fermentable sugars that yeast converts into alcohol. The type of honey used can significantly affect the flavor and color of the final product. Common types include clover, wildflower, and orange blossom honey, each contributing distinct flavor notes.
2. **Water**: Water is mixed with honey to create the must, which is the liquid that will be fermented. The quality of the water is crucial, as impurities or high mineral content can impact the taste and fermentation process. Ideally, use filtered or spring water.
3. **Yeast**: Yeast is the microorganism responsible for fermenting the sugars in honey into alcohol and carbon dioxide. While wild yeast can be used

for a traditional approach, using a cultivated yeast strain provides more control over the fermentation process and can help achieve consistent results.

Preparation Method

Creating the Must:

- Begin by measuring out honey and water. A typical ratio is 1 part honey to 4 parts water, though this can be adjusted based on desired sweetness and strength.
- In a large pot, heat the water to about 100°F (37°C) to help dissolve the honey. Avoid boiling, as excessive heat can alter the honey's delicate flavors.
- Stir in the honey until fully dissolved. This mixture, now called the must, should be clear and free of any undissolved honey.

Cooling and Pitching the Yeast:

- Allow the must to cool to room temperature, ideally around 70°F (21°C). This is important to prevent killing the yeast with excessive heat.
- Once cooled, transfer the must into a sanitized fermentation vessel, such as a glass carboy or a food-grade plastic bucket.
- Add the yeast to the must. If using dry yeast, sprinkle it directly onto the surface of the must. If using liquid yeast, pour it in gently. There is no need to stir; the yeast will naturally mix with the must.

Fermentation:

- Seal the fermentation vessel with an airlock or a cloth to allow gases to escape while preventing contaminants from entering.
- Place the vessel in a dark, cool area with a consistent temperature, ideally between 65-75°F (18-24°C).

- Over the next few days to weeks, you should observe bubbling in the airlock or see sediment forming at the bottom of the vessel, indicating active fermentation.

Racking and Aging:

- After fermentation slows down (usually after a few weeks), the mead should be transferred, or "racked," to a new, sanitized vessel. This helps separate the mead from the sediment and ensures a clearer final product.
- Seal the new vessel and allow the mead to age for several weeks to several months. Aging improves the flavor and allows the mead to mellow. The aging process can vary based on personal preference and desired flavor profile.

Bottling:

- Once aging is complete, the mead can be bottled. Ensure that bottles and caps are thoroughly sanitized to prevent contamination.
- Carefully siphon the mead into bottles, leaving some headspace at the top to accommodate any potential carbonation. Seal the bottles with caps or corks.

Conditioning:

- Allow the bottled mead to condition for a few more weeks. This step helps the flavors to integrate and can enhance the overall taste.

Cultural Significance

Basic traditional mead has been enjoyed across various cultures for thousands of years. Its simple ingredients and method reflect early brewing practices and the ingenuity of ancient brewers. Mead was often consumed during significant events, feasts, and ceremonies, symbolizing prosperity and community. In many cultures, it held a sacred or celebratory role, linking it to divine or heroic figures.

In modern times, traditional mead continues to be celebrated for its historical roots and versatility. It offers a connection to past brewing traditions and allows enthusiasts to experience a beverage that has stood the test of time.

Basic traditional mead is a testament to the simplicity and elegance of ancient brewing. With its core ingredients of honey, water, and yeast, it represents a fundamental approach to fermentation that has been cherished across cultures for centuries.

Orange Blossom Mead

Orange Blossom Mead is a type of traditional mead that uses orange blossom honey as its primary sweetener, imparting a distinctive floral and citrus flavor to the final product. This mead is appreciated for its delicate and aromatic profile, making it a popular choice among mead enthusiasts.

Ingredients

1. **Orange Blossom Honey**: This honey is collected from the nectar of orange blossoms, typically from orange trees. It has a light, floral flavor with subtle citrus notes, which can enhance the overall flavor profile of the mead. The quality and freshness of the honey are crucial for achieving the desired taste.
2. **Water**: The water used in Orange Blossom Mead should be clean and free from impurities. Filtered or spring water is ideal as it ensures that no off-flavors affect the mead.

3. **Yeast**: The choice of yeast can influence the final flavor and aroma of the mead. For Orange Blossom Mead, a neutral or mildly fruity yeast strain is often preferred to complement the honey's delicate flavors. Common choices include wine yeast strains such as Lalvin D-47 or Wyeast 4184.

Preparation Method

Creating the Must:

- **Measure and Mix**: Start by measuring out the orange blossom honey and water. A common ratio is 1 part honey to 4 parts water, but this can be adjusted based on the desired sweetness and strength of the mead.
- **Heat and Dissolve**: In a large pot, gently heat the water to around 100°F (37°C). Avoid boiling to preserve the honey's delicate flavors. Stir in the honey until it is fully dissolved, creating a clear and homogeneous mixture known as the must.

Cooling and Pitching the Yeast:

- **Cool the Must**: Allow the honey-water mixture to cool to room temperature, approximately 70°F (21°C). This is crucial as high temperatures can kill the yeast.
- **Add the Yeast**: Once cooled, transfer the must into a sanitized fermentation vessel. Add the yeast according to the manufacturer's instructions. If using dry yeast, sprinkle it directly onto the surface of the must. If using liquid yeast, pour it in gently. Stirring is not necessary, as the yeast will naturally mix with the must.

Fermentation:

- **Seal and Ferment**: Seal the fermentation vessel with an airlock to

allow gases to escape while preventing contamination. Place the vessel in a dark, cool area with a stable temperature, ideally between 65-75°F (18-24°C).
- **Monitor Fermentation**: Over the next few days to weeks, monitor the fermentation process. Look for signs such as bubbling in the airlock or sediment forming at the bottom of the vessel. This indicates that the yeast is actively converting the sugars into alcohol.

Racking and Aging:

- **Rack the Mead**: After fermentation has slowed or stopped, transfer the mead to a new, sanitized vessel. This process, known as racking, helps separate the mead from the sediment and clarifies the liquid.
- **Age the Mead**: Seal the new vessel and allow the mead to age. Aging can last several weeks to several months. This period allows the flavors to meld and develop, resulting in a smoother and more refined taste.

Bottling:

- **Prepare Bottles**: Ensure that bottles and caps are thoroughly sanitized to avoid contamination. Use bottles suitable for mead, such as wine bottles with corks or capped beer bottles.
- **Bottle the Mead**: Carefully siphon the aged mead into bottles, leaving a small amount of headspace at the top. Seal the bottles with caps or corks.

Conditioning:

- **Condition the Bottles**: Allow the bottled mead to condition for additional weeks. This step helps the flavors to integrate fully and can enhance the overall taste of the mead.

Characteristics and Flavor Profile

Orange Blossom Mead is known for its light, floral, and citrusy flavor profile. The orange blossom honey imparts a subtle yet distinctive sweetness and aroma, with hints of orange and a delicate floral note. The mead typically exhibits a pale golden color, which reflects the light color of the honey.

The flavor of Orange Blossom Mead can be influenced by several factors, including the quality of the honey, the yeast strain used, and the aging process. A well-made Orange Blossom Mead should balance the honey's sweetness with a slight acidity, offering a refreshing and aromatic drinking experience.

Cultural Significance

Orange Blossom Mead, while not as historically prominent as some other traditional meads, reflects a modern appreciation for the versatility of mead-making. The use of orange blossom honey ties the beverage to the regions where orange trees are cultivated, such as Florida and California, highlighting local agricultural products and preferences.

Acacia Honey Mead

Acacia Honey Mead is a distinctive type of mead that uses acacia honey as its primary sweetener. Acacia honey, known for its light color and mild, delicate flavor, imparts unique characteristics to the mead, making it a favored choice among enthusiasts.

Ingredients

1. **Acacia Honey**: Acacia honey is derived from the nectar of acacia tree blossoms, often from species such as black locust or Robinia pseudoacacia. It is characterized by its pale color, mild flavor, and high fructose content, which helps it remain liquid longer than other types of honey. This honey's subtle floral notes and light taste are key to creating a refined mead.
2. **Water**: The quality of water used in mead-making is crucial. For Acacia Honey Mead, it is best to use filtered or spring water to avoid any impurities or chlorine that might affect the taste or fermentation process.
3. **Yeast**: The choice of yeast impacts the flavor and fermentation of the mead. For Acacia Honey Mead, a neutral or mild yeast strain is ideal to preserve the delicate flavors of the honey. Popular options include wine yeast strains like Lalvin K1-V1116 or EC-1118.

Preparation Method

Creating the Must:

- **Measure and Mix**: Start by measuring the acacia honey and water. A typical ratio is 1 part honey to 4 parts water, but this can be adjusted depending on desired sweetness and mead strength.
- **Heat and Dissolve**: In a large pot, gently heat the water to approximately 100°F (37°C). Avoid boiling, as high temperatures can alter the delicate flavors of the honey. Stir in the acacia honey until it is completely dissolved, creating a clear, homogeneous mixture called the must.

Cooling and Pitching the Yeast:

- **Cool the Must**: Allow the honey-water mixture to cool to room temperature, ideally around 70°F (21°C). This prevents killing the yeast

with excessive heat.
- **Add the Yeast**: Transfer the cooled must into a sanitized fermentation vessel. Add the yeast according to the manufacturer's instructions. If using dry yeast, sprinkle it directly onto the surface of the must. For liquid yeast, pour it in gently. There is no need to stir, as the yeast will naturally integrate with the must.

Fermentation:

- **Seal and Ferment**: Seal the fermentation vessel with an airlock to allow gases to escape while preventing contaminants from entering. Place the vessel in a dark, cool area with a stable temperature, ideally between 65-75°F (18-24°C).
- **Monitor Fermentation**: Over the next few days to weeks, monitor the fermentation process. Signs of active fermentation include bubbling in the airlock and sediment forming at the bottom of the vessel. This indicates that the yeast is converting the honey's sugars into alcohol.

Racking and Aging:

- **Rack the Mead**: Once fermentation has slowed or stopped (usually after a few weeks), transfer the mead to a new, sanitized vessel to separate it from the sediment. This process, known as racking, helps clarify the mead and improve its taste.
- **Age the Mead**: Seal the new vessel and allow the mead to age. Aging can last from several weeks to several months. This period allows the flavors to develop and integrate, resulting in a smoother and more refined mead.

Bottling:

- **Prepare Bottles**: Ensure that bottles and caps are thoroughly sanitized to prevent contamination. Use bottles suitable for mead, such as wine

bottles with corks or capped beer bottles.
- **Bottle the Mead**: Carefully siphon the aged mead into bottles, leaving a small amount of headspace at the top. Seal the bottles with caps or corks to prevent oxidation.

Conditioning:

- **Condition the Bottles**: Allow the bottled mead to condition for a few more weeks. This additional time helps the flavors to meld and can enhance the overall taste.

Characteristics and Flavor Profile

Acacia Honey Mead is known for its light, subtle flavor profile. The acacia honey imparts a delicate floral and slightly fruity taste with a hint of vanilla. The mead typically exhibits a pale, almost transparent color, reflecting the light hue of the honey. The flavor of Acacia Honey Mead is gentle and smooth, with a clean finish that highlights the honey's natural sweetness without being overpowering.

The high fructose content of acacia honey helps the mead retain a pleasant sweetness even after fermentation, while the moderate acidity balances the sweetness and adds complexity. The overall taste experience is refined and elegant, making it an excellent choice for those who appreciate a milder, more nuanced mead.

Cultural Significance

While Acacia Honey Mead may not have the same historical prominence as some other traditional meads, it reflects a modern appreciation for diverse honey varieties and their impact on mead flavor. The use of acacia honey highlights the regional cultivation of acacia trees and the honey's unique properties.

CHAPTER FIFTEEN

In contemporary mead-making, Acacia Honey Mead represents a sophisticated choice for those looking to explore different flavor profiles and experiment with the subtle nuances of various honeys. It showcases the versatility of mead as a beverage and its ability to adapt to different ingredients and preferences.

16

CHAPTER SIXTEEN

FRUITS AND SPICE MEADS

Mead, often hailed as the "drink of the gods," can be crafted from a variety of ingredients, but it's the fruit and spice meads that truly stole my heart. They represent a fusion of tradition and innovation, allowing me to explore a world of flavors that is as rich as it is diverse.

Fruit meads, or "melomels," are a beautiful showcase of how fruit can elevate the humble mead. My journey into fruit meads began with a batch of apple mead. The aroma of fresh apples combined with the honey's sweet undertones was nothing short of intoxicating. I remember the excitement of pressing the apples, the juice mingling with honey, and the anticipation of what this combination would become.

In crafting fruit meads, the choice of fruit is paramount. Apples, berries, peaches, and even exotic fruits like mangoes or pineapples can transform a mead. I experimented with a raspberry mead one summer, using a bounty of freshly picked berries. The vibrant red color and tartness of the raspberries created a contrast with the honey's sweetness, resulting in a mead that was both refreshing and complex. The fruit's natural acidity helped balance the sweetness, adding layers of flavor that delighted my palate.

One of the most rewarding aspects of fruit meads is the opportunity to blend different fruits and experiment with ratios. A pear and ginger mead I once created was a revelation. The pear added a delicate sweetness, while the ginger provided a warm, spicy kick. The interplay of flavors was a testament to how fruits can harmonize with honey, creating a drink that is not just enjoyable but also uniquely personal.

Spice meads, or "metheglins," introduce a different kind of magic to the mead-making process. The idea of adding spices like cinnamon, cloves, or even more exotic spices such as cardamom and star anise intrigued me. I recall my first foray into spice meads with a chai spiced mead. Inspired by the rich, aromatic flavors of chai tea, I combined cinnamon, cloves, and a hint of black pepper with honey.

The result was a mead that felt both comforting and sophisticated. The spices didn't just add flavor; they created an experience. Each sip was like a warm embrace, with the honey providing a sweet base that allowed the spices to shine. This mead became a favorite during the colder months, offering a flavorful escape from the chill.

Experimenting with spices also meant paying attention to their potency. I learned early on that a little goes a long way. A batch of mead I made with too much cinnamon resulted in an overpowering flavor that overshadowed the honey. Finding the right balance was crucial, and it became clear that the art of making spice meads was as much about precision as it was about creativity.

My most memorable creation was a peach and cardamom mead. The peaches added a lush, fruity sweetness, while the cardamom introduced an exotic spice that was both aromatic and warm. The resulting mead was a symphony of flavors that danced on the tongue, each sip revealing new notes.

Blending fruit and spice requires a delicate touch. It's essential to ensure that neither the fruit nor the spice overwhelms the mead. I found that starting with small amounts of spices and adjusting as the fermentation progressed helped maintain balance. Tasting periodically allowed me to fine-tune the flavor profile, ensuring that the final product was harmonious and delightful.

Blackberry Melomel

I've found that lighter honey varieties, like clover or orange blossom, complement the bright, fruity notes of blackberries without overpowering them.

Blackberries, on the other hand, are known for their deep, complex flavor. Their natural sweetness is balanced by a hint of tartness, which adds depth to the melomel. The process begins with selecting ripe, high-quality blackberries. Fresh or frozen blackberries both work well, but I've found that using fresh berries when they're in season results in a more vibrant flavor.

The Brewing Process

The creation of a blackberry melomel involves several key steps. The first is making the must, which is a mixture of honey, water, and blackberries. I usually start by heating the water slightly and dissolving the honey into it. This ensures that the honey is fully integrated and helps create a more uniform must.

Once the honey is fully incorporated, I add the blackberries. Crushing the berries or blending them helps release their juice, enhancing the flavor infusion. In my experience, adding the berries directly to the must and

CHAPTER SIXTEEN

allowing them to ferment with the mixture yields a more robust blackberry character. I often let this mixture sit for a day or two, allowing the flavors to meld before adding yeast.

The yeast selection can influence the final flavor profile of the melomel. A neutral wine yeast works well to let the blackberry flavor shine, while some brewers prefer a fruit-specific yeast to complement the berries. After pitching the yeast, I let the must ferment in a controlled environment, monitoring the temperature and fermentation progress.

Aging and Maturation

One of the joys of making blackberry melomel is the aging process. After fermentation, the melomel benefits from aging, which helps mellow out the flavors and integrate the honey and blackberry notes. I usually transfer the mead to secondary fermentation vessels and let it age for several months. This maturation period allows the blackberry flavor to develop further, becoming more pronounced and complex.

During this time, I also monitor the melomel for clarity and taste. Occasionally, I might need to rack it to a new vessel to remove sediment or adjust the flavor profile if necessary. The aging process is where the mead truly transforms, becoming smoother and more refined.

Tasting Notes and Pairings

When the blackberry melomel is finally ready, the result is a beautifully balanced beverage. The color is often a deep purple, reflecting the richness of the blackberries. On the nose, the melomel offers an inviting aroma of ripe blackberries and honey. The taste is a harmonious blend of sweet and tart, with the honey providing a mellow, smooth backdrop to the vibrant fruit flavors.

I've found that blackberry melomel pairs wonderfully with a variety of foods. Its richness complements savory dishes like roasted pork or duck, while its fruity notes enhance desserts such as dark chocolate or cheesecake. It's also delightful on its own, enjoyed chilled or at room temperature, making it a versatile choice for various occasions.

Personal Reflections

Crafting a blackberry melomel has been one of the more satisfying experiences in my mead-making journey. The process of blending the honey with the bold flavors of blackberries creates a beverage that is both elegant and approachable. Every step, from selecting the ingredients to the final tasting, has been a lesson in patience and precision.

Hibiscus and Ginger Metheglin

This metheglin, a type of spiced mead, combines the vibrant tartness of hibiscus with the warming zing of ginger, resulting in a beverage that's both complex and invigorating. Here's a detailed exploration of hibiscus and ginger metheglin, including its recipe and brewing process.

The Ingredients

Honey: The base of any metheglin, including hibiscus and ginger, is honey. A medium to dark honey works well to provide a rich, robust flavor that complements the other ingredients. While lighter honey varieties can be used, they might not offer the same depth of flavor.

Hibiscus: Dried hibiscus flowers are the primary source of floral and tart notes in this metheglin. Hibiscus imparts a vibrant red color and a tangy flavor that pairs beautifully with the sweetness of honey.

Ginger: Fresh ginger adds a warm, spicy kick that balances the floral notes of hibiscus. Its pungent aroma and zesty flavor enhance the overall complexity of the mead.

Additional Ingredients: Depending on personal preference, you might also consider adding complementary spices such as cinnamon or cloves, which can add extra layers of flavor.

The Recipe

Ingredients:

- 3 pounds of honey (medium to dark)
- 1 gallon of water
- 1 cup dried hibiscus flowers
- 2-3 tablespoons freshly grated ginger (adjust to taste)
- 1 packet of wine yeast (such as a neutral wine yeast or a fruit-specific yeast)
- Optional: 1 cinnamon stick or 2-3 whole cloves
- Yeast nutrient (as per package instructions)

Equipment:

- Large pot
- Fermentation vessel (carboy or bucket with airlock)
- Sanitizing solution
- Siphon or racking cane
- Bottles and corks or caps

Brewing Process

Prepare the Must:

- In a large pot, heat about half a gallon of water and dissolve the honey

into it. Stir until fully mixed and the honey is completely dissolved. This mixture is known as the must.
- Once the honey is dissolved, add the remaining water to the pot. The total volume should be around 1 gallon.
- Add the dried hibiscus flowers and grated ginger to the must. If using, add the optional spices (cinnamon stick and/or cloves). Stir well and let the mixture steep for about 15-30 minutes. This will allow the flavors to infuse.

Cool and Strain:

- After steeping, cool the must to room temperature. Strain out the hibiscus flowers, ginger, and any spices using a fine mesh strainer or cheesecloth. This prevents any solid matter from affecting the fermentation process.

Pitch the Yeast:

- Transfer the strained must into a sanitized fermentation vessel. If you have a hydrometer, take a gravity reading to measure the specific gravity, which helps in estimating the final alcohol content.
- Rehydrate the yeast according to the package instructions and pitch it into the must. Add yeast nutrient if required.

Fermentation:

- Seal the fermentation vessel with an airlock and place it in a dark, temperature-controlled environment. The ideal temperature for fermentation is typically between 65-75°F (18-24°C).
- Allow the mead to ferment for about 2-4 weeks, or until fermentation activity subsides. You can check this by monitoring the airlock or taking periodic gravity readings.

Aging and Bottling:

- Once fermentation is complete, siphon the metheglin into a clean, sanitized secondary vessel, leaving behind any sediment. This process is called racking.
- Let the metheglin age for at least 1-2 months to allow the flavors to meld and develop. The aging process helps mellow out the sharp edges and enhances the overall complexity.
- After aging, bottle the metheglin in sanitized bottles and seal with corks or caps. Allow it to age further in the bottle if desired.

Tasting and Enjoying

The hibiscus and ginger metheglin should be a vibrant red with a complex flavor profile. The hibiscus imparts a tangy, floral note, while the ginger adds warmth and spice. The honey provides a sweet, smooth backdrop that balances the tartness of the hibiscus and the heat of the ginger.

When tasting, you'll find that the flavors have blended beautifully, creating a refreshing and intriguing mead. This metheglin pairs well with a variety of dishes, including spicy foods, roasted meats, or even as an accompaniment to cheese and fruit platters.

17

CHAPTER SEVENTEEN

EXPERIMENTAL MEAD

Experimental mead represents a dynamic evolution of the traditional craft of mead-making, known for its simplicity with just honey, water, and yeast. Modern brewers are innovating by introducing unconventional ingredients like exotic fruits, spices, and alternative fermentables, as well as employing advanced techniques such as barrel aging and wild fermentation. Technology plays a crucial role, with new equipment and data analysis helping refine these experiments. The experimental mead community thrives on collaboration and knowledge-sharing, with events and online platforms supporting innovation. Despite challenges in balancing flavor and quality, experimental mead continues to expand the possibilities of this ancient beverage, offering new and exciting flavors to enthusiasts.

Chapter Seventeen

Coffee and Vanilla Mead

Coffee and vanilla mead combines the complex, robust flavors of coffee with the sweet, aromatic essence of vanilla, creating a mead that's rich and indulgent. This mead is perfect for those who enjoy a touch of caffeine in their fermented beverages or simply want to explore unique flavor combinations in their home brewing.

Flavor Profile:

- **Coffee:** Adds a deep, roasted, and slightly bitter note, which contrasts beautifully with the sweetness of the honey.
- **Vanilla:** Contributes a smooth, creamy, and slightly floral sweetness, enhancing the overall richness of the mead.
- **Honey:** The primary fermentable sugar, it provides a natural sweetness and complexity that complements both the coffee and vanilla.

Recipe for Coffee and Vanilla Mead

Ingredients:

- **Honey:** 3 pounds (1.36 kg) – Use a high-quality, flavorful honey like orange blossom or clover for the best results.
- **Water:** 1 gallon (3.78 liters) – Preferably filtered or spring water to avoid chlorine or other impurities.
- **Coffee Beans:** 2 ounces (56 grams) – Coarsely ground. Choose a good quality coffee with a flavor profile you enjoy.
- **Vanilla Beans:** 2 – Split and scraped, or use 2 tablespoons of high-quality vanilla extract as a substitute.
- **Yeast:** 1 packet (5 grams) of mead or wine yeast – Lalvin D-47 or Wyeast 4184 Sweet Mead are good choices.

- **Yeast Nutrient:** 1 teaspoon – To ensure a healthy fermentation.
- **Yeast Energizer:** 1/2 teaspoon – Optional but helpful for improving yeast performance.

Equipment:

- 1-gallon fermenter (with airlock and stopper)
- Large pot
- Stirring spoon
- Fine mesh strainer or cheesecloth
- Sanitizing solution

Instructions

- **Sanitize:** Thoroughly clean and sanitize all equipment to prevent contamination.
- **Prepare Coffee:** Coarsely grind the coffee beans. You can cold brew the coffee for a less acidic flavor, or if you prefer, brew it hot and then let it cool. For cold brewing, steep the grounds in a small amount of cold water for 12-24 hours, then strain out the grounds. For hot brewing, brew a strong cup of coffee, let it cool to room temperature, and use it in the next step.

Make the Must:

- In a large pot, heat about 1/2 gallon (1.89 liters) of water to a boil. Remove from heat and dissolve the honey in the hot water, stirring well.
- Add the remaining water to the pot to cool it down to room temperature.

Add Coffee and Vanilla:

- If using cold brewed coffee, add it directly to the honey-water mixture.
- If using hot brewed coffee, make sure it has cooled to room temperature

before adding it to the honey-water mixture.
- Split the vanilla beans and scrape the seeds into the mixture. Alternatively, add vanilla extract.

Add Yeast and Nutrients:

- Once the mixture has cooled to room temperature, sprinkle the yeast on top. Allow it to sit for 15 minutes before gently stirring it in.
- Add yeast nutrient and yeast energizer (if using), stirring gently to mix.

Fermentation:

- Pour the mixture into your sanitized fermenter, leaving some headspace.
- Attach the airlock and stopper to the fermenter.
- Store the fermenter in a dark, cool place (around 65-75°F or 18-24°C) for primary fermentation.

Primary Fermentation:

- Allow the mead to ferment for 2-4 weeks, or until the fermentation activity slows down. This can be checked by observing the airlock or taking gravity readings with a hydrometer.

Secondary Fermentation:

- Once primary fermentation is complete, carefully rack (transfer) the mead into a clean, sanitized secondary fermenter, leaving sediment behind.
- Age the mead for 1-3 months to develop flavors and clarity.

Bottling:

- After aging, the mead can be bottled. Sanitize bottles and caps.

- Transfer the mead into bottles, leaving some headspace, and seal with caps or corks.

Maturation:

- Let the bottled mead age for at least another month. The flavors will continue to meld and develop.

Notes:

- **Adjusting Coffee Flavor:** If you prefer a stronger coffee flavor, consider adding more coffee or adjusting the brewing method. For a milder flavor, use less coffee or reduce the steeping time.
- **Vanilla Intensity:** For a more pronounced vanilla taste, you can increase the number of vanilla beans or use additional vanilla extract, but do so carefully to avoid overpowering the mead.

This Coffee and Vanilla Mead offers a sophisticated twist on traditional mead, perfect for sipping on a cool evening or sharing with friends.

Chocolate Cherry Bochet

Chocolate Cherry Bochet is a rich, dessert-like mead that combines the deep flavors of caramelized honey with the indulgent taste of chocolate and the sweet tartness of cherries. Bochet, a style of mead where honey is caramelized before fermentation, adds an extra layer of complexity and richness to the flavor profile.

Flavor Profile:

- **Caramelized Honey:** Provides a deep, rich, and slightly burnt sugar

flavor, enhancing the overall complexity of the mead.
- **Chocolate:** Adds a creamy, bittersweet chocolate essence that pairs beautifully with the caramelized honey.
- **Cherries:** Contribute a fruity, tart flavor that complements the richness of the chocolate and honey.

Recipe for Chocolate Cherry Bochet

Ingredients:

- **Honey:** 4 pounds (1.81 kg) – A high-quality honey such as wildflower or clover.
- **Water:** 1 gallon (3.78 liters) – Filtered or spring water is preferable.
- **Cherries:** 2 pounds (900 grams) – Fresh or frozen, pitted and chopped.
- **Cocoa Powder:** 1 cup (100 grams) – Unsweetened cocoa powder or cacao for a richer chocolate flavor.
- **Vanilla Extract:** 1 tablespoon – To enhance the chocolate and cherry flavors.
- **Yeast:** 1 packet (5 grams) of mead or wine yeast – Lalvin D-47 or Wyeast 4184 Sweet Mead are good options.
- **Yeast Nutrient:** 1 teaspoon – To promote healthy fermentation.
- **Yeast Energizer:** 1/2 teaspoon – Optional, for better yeast performance.

Equipment:

- 1-gallon fermenter (with airlock and stopper)
- Large pot
- Stirring spoon
- Fine mesh strainer or cheesecloth
- Sanitizing solution

Instructions:

1. **Sanitize:** Clean and sanitize all equipment thoroughly to avoid contamination.
2. **Caramelize the Honey:**

- In a large pot, heat the honey over medium heat, stirring constantly until it begins to darken and develop a caramel-like color and aroma. This process should take about 15-20 minutes. Be careful not to burn the honey.

Prepare the Must:

- Gradually add water to the caramelized honey while stirring to help dissolve it and cool the mixture. Continue stirring until the mixture reaches room temperature.
- Add the cocoa powder to the honey-water mixture, stirring well to ensure it is fully dissolved.

Add Cherries and Vanilla:

- If using fresh cherries, pit and chop them before adding. If using frozen cherries, thaw and chop them. Add the cherries to the honey mixture.
- Stir in the vanilla extract.

Add Yeast and Nutrients:

- Once the mixture has cooled to room temperature, sprinkle the yeast on top. Let it sit for 15 minutes, then stir gently to incorporate.
- Add yeast nutrient and yeast energizer (if using), stirring well.

Fermentation:

- Pour the mixture into the sanitized fermenter, leaving some headspace.
- Attach the airlock and stopper.
- Store the fermenter in a dark, cool place (around 65-75°F or 18-24°C) for primary fermentation.

Primary Fermentation:

- Allow the mead to ferment for 2-4 weeks, or until fermentation activity slows. You can check this by observing the airlock or taking gravity readings with a hydrometer.

Secondary Fermentation:

- Once primary fermentation is complete, carefully rack (transfer) the mead into a clean, sanitized secondary fermenter, leaving sediment behind.
- Age the mead for 2-3 months to develop the flavors and clarity.

Bottling:

- After aging, sanitize bottles and caps.
- Transfer the mead into bottles, leaving some headspace, and seal with caps or corks.

Maturation:

- Allow the bottled mead to age for at least another month. The flavors will continue to meld and develop, resulting in a smooth, flavorful mead.

Notes:

- **Adjusting Chocolate Flavor:** If you prefer a stronger chocolate flavor, you can increase the amount of cocoa powder. For a subtler chocolate

note, reduce the cocoa powder slightly.
- **Cherry Intensity:** To adjust the cherry flavor, you can add more cherries or use cherry juice concentrate if you want a more pronounced cherry taste.
- **Sweetness Level:** If you find the bochet too sweet or not sweet enough, you can adjust the sweetness by adding a small amount of honey or sweetener after fermentation, but do so carefully to avoid over-sweetening.

Chocolate Cherry Bochet offers a rich, indulgent experience that combines the caramelized complexity of bochet with the creamy sweetness of chocolate and the fruity tartness of cherries.

Lavender Lemon Balm Mead

Lavender Lemon Balm Mead is a refreshing and aromatic mead that blends the calming notes of lavender with the invigorating taste of lemon balm. This mead is perfect for those who appreciate herbal and floral flavors and are looking for a unique twist on traditional mead.

Flavor Profile:

- **Lavender:** Provides a fragrant, floral, and slightly sweet aroma that adds a soothing quality to the mead.
- **Lemon Balm:** Adds a bright, citrusy flavor with a hint of mint, enhancing the freshness and complexity of the mead.
- **Honey:** Offers a natural sweetness that balances the herbal notes and adds depth to the flavor.

Recipe for Lavender Lemon Balm Mead

Ingredients:

- **Honey:** 3 pounds (1.36 kg) – Choose a mild honey like clover or acacia to allow the herbal flavors to shine.
- **Water:** 1 gallon (3.78 liters) – Filtered or spring water is ideal to avoid any off-flavors from chlorine.
- **Dried Lavender:** 1/2 cup (12 grams) – Ensure it's culinary-grade for the best flavor.
- **Fresh Lemon Balm Leaves:** 1 cup (30 grams) – Lightly crushed to release their essential oils. If using dried lemon balm, use 1/4 cup (6 grams).
- **Yeast:** 1 packet (5 grams) of mead or wine yeast – Lalvin D-47 or Wyeast 4184 Sweet Mead are good choices.
- **Yeast Nutrient:** 1 teaspoon – To support a healthy fermentation.
- **Yeast Energizer:** 1/2 teaspoon – Optional, for enhanced yeast performance.

Equipment:

- 1-gallon fermenter (with airlock and stopper)
- Large pot
- Stirring spoon
- Fine mesh strainer or cheesecloth
- Sanitizing solution

Instructions:

Sanitize

Thoroughly clean and sanitize all equipment to prevent contamination.

Prepare the Must

- In a large pot, heat about 1/2 gallon (1.89 liters) of water until it just begins to boil. Remove from heat.
- Add the honey to the hot water, stirring until fully dissolved. This creates a honey-water mixture known as the must.
- Add the remaining water to the pot to cool the mixture to room temperature.

Infuse Lavender and Lemon Balm

- Place the dried lavender and fresh lemon balm leaves (or dried lemon balm) in a large heat-safe container.
- Pour the cooled honey-water mixture over the herbs.
- Cover and steep for 15-30 minutes, depending on how strong you want the herbal flavors to be. Longer steeping times will yield more intense flavors.

Strain the Must

- After steeping, strain out the herbs using a fine mesh strainer or cheesecloth to remove any solid particles.

Add Yeast and Nutrients

- Once the mixture has cooled to room temperature, sprinkle the yeast on top. Let it sit for 15 minutes, then stir gently to incorporate.
- Add yeast nutrient and yeast energizer (if using), stirring well.

Fermentation

- Pour the mixture into your sanitized fermenter, leaving some headspace.
- Attach the airlock and stopper.
- Store the fermenter in a dark, cool place (around 65-75°F or 18-24°C) for primary fermentation.

Primary Fermentation

- Allow the mead to ferment for 2-4 weeks, or until the fermentation activity slows down. Check this by observing the airlock or taking gravity readings with a hydrometer.

Secondary Fermentation

- Once primary fermentation is complete, carefully rack (transfer) the mead into a clean, sanitized secondary fermenter, leaving sediment behind.
- Age the mead for 1-2 months to develop flavors and clarity.

Bottling

- After aging, sanitize bottles and caps.
- Transfer the mead into bottles, leaving some headspace, and seal with caps or corks.

Maturation

- Allow the bottled mead to age for at least another month. This will allow the herbal flavors to mellow and integrate, resulting in a smooth and balanced mead.

Notes:

- **Herbal Intensity:** Adjust the amount of lavender and lemon balm to match your flavor preference. For a more subtle herbal flavor, reduce the quantity; for a more pronounced taste, increase it.
- **Sweetness Level:** If you prefer a sweeter mead, you can add a small amount of honey after fermentation, but do so gradually and taste frequently to avoid over-sweetening.

- **Additional Flavors:** Consider experimenting with other herbs or spices that complement the lavender and lemon balm, such as a touch of mint or chamomile, for a unique twist.

Lavender Lemon Balm Mead offers a beautifully aromatic and refreshing experience, perfect for sipping on a warm day or serving at a garden party.

18

CHAPTER EIGHTEEN

SEASONAL AND HOLIDAY MEADS

Seasonal and holiday meads capture the essence of different times of the year with unique flavors:

- **Winter**: Spiced with cinnamon, cloves, and nutmeg for warmth.
- **Spring**: Light and floral, featuring elderflowers, chamomile, or fresh berries.
- **Summer**: Fruity and refreshing with ingredients like peaches or tropical fruits.
- **Autumn**: Earthy and cozy, incorporating spices like cinnamon and nutmeg, along with apples or pumpkins.
- **Holiday**: Special meads tailored to specific celebrations, such as spiced Christmas meads or cranberry meads for Thanksgiving.

These meads enhance seasonal and festive occasions with their distinctive flavors.

Autumn Spice Mead

The primary allure of autumn spice mead lies in its ability to evoke the cozy, inviting atmosphere of fall. This mead typically features a harmonious blend of spices such as cinnamon, nutmeg, cloves, and allspice, reminiscent of seasonal treats like pumpkin pie and spiced cider. These spices are carefully balanced to complement the natural sweetness of honey, resulting in a mead that is both aromatic and flavorful.

In addition to the spices, autumn spice mead often incorporates other elements of the season, such as apples or pears, which add depth and complexity to the flavor profile. The choice of honey also plays a crucial role; darker honeys with robust flavors, such as buckwheat or chestnut honey, are particularly well-suited for this spiced variant, as they can stand up to and complement the strong spices.

The Recipe

Here's a straightforward recipe to create your own autumn spice mead:

Ingredients:

- 3 pounds (1.36 kg) of honey (preferably dark, such as buckwheat or chestnut honey)
- 1 gallon (3.78 liters) of water (filtered or distilled)
- 1 packet of mead yeast (such as Lalvin D-47 or Wyeast 4184)
- 1 teaspoon of yeast nutrient
- 1 teaspoon of yeast energizer
- 2-3 cinnamon sticks
- 5-6 whole cloves
- 1/2 teaspoon of ground nutmeg
- 1/2 teaspoon of allspice
- Optional: 1-2 cups of apple juice or pear juice
- Optional: 1-2 cups of chopped apples or pears

Instructions:

Prepare the Must:

- Heat about half of the water in a large pot until it reaches a simmer. Add the honey, stirring until it is fully dissolved.
- Once the honey is dissolved, remove the pot from heat and add the remaining water to cool the mixture. The goal is to bring the temperature down to room temperature or slightly below to prevent harming the yeast.

Add Spices and Flavorings:

- If using, add the chopped apples or pears to the cooled honey-water mixture.
- Stir in the cinnamon sticks, cloves, nutmeg, and allspice. These spices will infuse their flavors into the mead during fermentation.

Prepare the Yeast:

- Rehydrate the yeast according to the manufacturer's instructions and add it to the must. This can be done by sprinkling the yeast over the surface and letting it sit for about 15 minutes, then gently stirring.

Add Nutrients and Energizers:

- Add the yeast nutrient and yeast energizer to the must. These additives help ensure a healthy fermentation by providing essential nutrients for the yeast.

Fermentation:

- Transfer the must to a sanitized fermentation vessel, such as a glass

carboy or plastic fermenter, and seal it with an airlock.
- Allow the mead to ferment in a dark, cool place at a temperature range of 60-70°F (15-21°C) for about 2-4 weeks, or until the fermentation activity subsides.

Secondary Fermentation and Aging:

- After primary fermentation, siphon the mead into a clean, sanitized secondary fermenter, leaving behind the sediment.
- Age the mead for an additional 2-3 months to allow the flavors to meld and the spices to integrate fully.

Bottling and Enjoying:

- Once aging is complete, bottle the mead in clean, sanitized bottles and seal them with caps or corks.
- Allow the bottled mead to age for another month or so before tasting. This additional aging helps to refine the flavors and improve the overall balance.

Winter Solstice Mulled Mead

Mulled mead is essentially a spiced and heated version of traditional mead. The Winter Solstice Mulled Mead takes inspiration from classic mulled wines and ciders, incorporating seasonal spices and ingredients that evoke warmth and festivity. The Winter Solstice, marking the shortest day of the year and the official beginning of winter, serves as a fitting backdrop for this rich and aromatic beverage.

The flavor profile of Winter Solstice Mulled Mead typically features a blend of spices such as cinnamon, cloves, and star anise, combined with the sweetness of honey and the tartness of winter fruits like oranges and cranberries. This combination not only enhances the natural flavors of the mead but also creates a drink that is both refreshing and soothing during the colder months.

The Recipe

Here's a simple yet flavorful recipe to create your own Winter Solstice Mulled Mead:

Ingredients:

- 1 bottle (750 ml) of mead (preferably a semi-sweet or sweet mead)
- 1 orange, sliced
- 1 apple, sliced
- 1/2 cup of fresh cranberries
- 2-3 cinnamon sticks
- 4-5 whole cloves
- 2-3 star anise pods
- 1/2 teaspoon of ground ginger
- 1/4 cup of honey (if needed, to adjust sweetness)
- Optional: 1/4 cup of brandy or orange liqueur (for added depth)

Instructions

Prepare the Ingredients:

- Wash and slice the orange and apple into thin rounds. If using, lightly crush the cranberries to release their juices.

Heat the Mead:

- Pour the mead into a large pot or saucepan and heat it over medium-low heat. Avoid boiling to prevent evaporating the alcohol and altering the flavors.

Add the Spices and Fruits:

- Add the orange slices, apple slices, cranberries, cinnamon sticks, cloves, star anise, and ground ginger to the heating mead. Stir gently to combine.

Simmer:

- Allow the mixture to simmer for about 20-30 minutes. This process will enable the spices and fruits to infuse their flavors into the mead. Stir occasionally and taste to adjust the sweetness.

Adjust Sweetness (Optional):

- Taste the mulled mead and add honey if a sweeter flavor is desired. Stir well to ensure the honey is fully incorporated.

Add Brandy or Liqueur (Optional):

- For an extra touch of warmth and depth, stir in the brandy or orange liqueur during the last few minutes of simmering. This step is optional but can enhance the overall flavor.

Serve:

- Strain the mulled mead to remove the solid spices and fruit pieces before serving.
- Pour the warm mead into heatproof mugs or glasses and serve immediately.

CHAPTER EIGHTEEN

Spring Blossom Mead

Spring Blossom Mead is characterized by its light and fragrant profile, showcasing the flavors of spring flowers and fresh fruits. The mead typically features floral ingredients such as elderflowers, hibiscus, or chamomile, which contribute to its aromatic qualities. Additionally, fruits like apples, pears, and berries may be included to enhance the flavor profile and provide a refreshing contrast to the floral notes.

The choice of honey also plays a significant role in crafting Spring Blossom Mead. Lighter honeys, such as clover or acacia, are often used to complement the delicate flavors and ensure that the floral and fruity notes shine through without being overshadowed by the sweetness of the honey.

The Recipe

Here's a simple and elegant recipe to create your own Spring Blossom Mead:

Ingredients:

- 3 pounds (1.36 kg) of honey (preferably a light honey, such as clover or acacia)
- 1 gallon (3.78 liters) of water (filtered or distilled)
- 1 packet of mead yeast (such as Wyeast 4184 or Lalvin D-47)
- 1 teaspoon of yeast nutrient
- 1 teaspoon of yeast energizer
- 1/4 cup of dried elderflowers or 1/2 cup of fresh elderflowers
- 1/4 cup of dried hibiscus flowers (optional, for added color and flavor)
- 1-2 cups of chopped fresh fruit (such as apples, pears, or berries)
- Optional: 1/4 cup of fresh lemon juice (for added brightness)

Instructions

Prepare the Must:

- Heat about half of the water in a large pot until it reaches a simmer. Add the honey, stirring until it is fully dissolved.
- Once the honey is dissolved, remove the pot from heat and add the remaining water to cool the mixture to room temperature or slightly below.

Add Flowers and Fruit:

- If using, add the dried elderflowers, dried hibiscus flowers, and chopped fresh fruit to the cooled honey-water mixture.
- For a brighter flavor, you can also add fresh lemon juice at this stage.

Prepare the Yeast:

- Rehydrate the yeast according to the manufacturer's instructions and add it to the must. This is done by sprinkling the yeast over the surface and letting it sit for about 15 minutes before gently stirring.

Add Nutrients and Energizers:

- Stir in the yeast nutrient and yeast energizer to support healthy fermentation and ensure that the yeast has the necessary nutrients to thrive.

Fermentation:

- Transfer the must to a sanitized fermentation vessel, such as a glass carboy or plastic fermenter, and seal it with an airlock.
- Allow the mead to ferment in a cool, dark place at a temperature range of 60-70°F (15-21°C) for about 2-4 weeks, or until fermentation activity slows down.

Secondary Fermentation and Aging:

- After primary fermentation is complete, siphon the mead into a clean, sanitized secondary fermenter, leaving behind any sediment.
- Age the mead for an additional 2-3 months to allow the flavors to mature and meld.

Bottling and Enjoying:

- Once the aging process is complete, bottle the mead in clean, sanitized bottles and seal them with caps or corks.
- Allow the bottled mead to age for another month or so before tasting to allow the flavors to fully develop.

19

CHAPTER NINETEEN

MEAD PAIRING AND COCKTAILS

Mead, a versatile and ancient beverage, offers exciting possibilities for food pairings and cocktails.

Food Pairings:

- **Sweet Meads**: Pair with rich desserts or spicy dishes to balance sweetness and heat.
- **Dry Meads**: Complement light fare such as seafood, salads, or tangy cheeses.
- **Spiced Meads**: Match with hearty, autumnal dishes and baked goods like gingerbread.
- **Fruit-Infused Meads**: Enhance with corresponding fruits or desserts.

Cocktails:

- **Mead Mule**: Mix mead with ginger beer and lime for a honeyed twist on the Moscow Mule.

CHAPTER NINETEEN

- **Mead Sangria**: Combine mead with chopped fruits, brandy, and honey for a fruity, refreshing drink.
- **Spiced Mead Old Fashioned**: Blend spiced mead with bitters and sugar for a sophisticated cocktail.
- **Mead Mojito**: Muddle mint and lime with mead and soda water for a honeyed Mojito.
- **Mead and Tonic**: Simple and crisp, mix mead with tonic water and garnish with lime.

These pairings and cocktails highlight mead's diverse flavors and enhance its enjoyment in various culinary and drink settings.

Pairing Mead with Cheese, Meat, and Dessert

As someone who has always been intrigued by mead, its ability to pair with different foods has fascinated me even more. Mead, with its honey base, offers a range of flavors that can elevate a meal, whether it's light and floral or rich and spiced. Over time, I've experimented with pairing mead alongside cheese, meat, and dessert, and the results have been nothing short of delightful. Let me walk you through my experience of pairing mead with these food categories in detail.

Pairing Mead with Cheese

Cheese and wine pairings are a well-known tradition, but mead and cheese? That's something that's often overlooked. I found that the earthy, creamy nature of cheese complements mead in a way that brings out hidden layers of flavor in both.

Dry Meads with Goat Cheese

The tangy, slightly acidic bite of goat cheese pairs beautifully with a crisp, dry mead. When I first tried this pairing, the mead's dryness cut through the creaminess of the cheese, while the honey undertones softened the goat cheese's sharpness. This balance made for a surprisingly refreshing experience. A light clover honey mead worked wonders here.

Sweet Meads with Blue Cheese

Pairing a rich, sweet mead with the intense saltiness of blue cheese was a revelation for me. The bold flavors of blue cheese demand something that can stand up to its pungency, and a sweet mead—something like an orange blossom mead—did the job perfectly. The honey sweetness mellowed the cheese's saltiness while enhancing its creamy texture.

Spiced Meads with Aged Cheddar

The sharp, nutty flavors of aged cheddar, when paired with a spiced mead, became a cozy, hearty combination. The warm spices in the mead, such as cinnamon and cloves, played off the cheddar's sharpness and richness. I couldn't help but imagine this as the perfect pairing for a chilly autumn evening by the fire.

Pairing Mead with Meat

Mead's versatility extends to meats, where it can act as both a complement and a contrast, depending on the style of mead and the preparation of the meat. Here's where I had the most fun experimenting.

Dry Meads with Roasted Chicken

One of my go-to pairings is a dry mead with simple roasted chicken. The mead's clean, crisp profile doesn't overpower the dish but enhances the natural flavors of the chicken. A light acacia honey mead with herbal notes pairs particularly well, especially when the chicken is seasoned with rosemary or thyme. The result is a delicate balance of flavors, with the mead's subtle sweetness lifting the savory notes of the roast.

Sweet Meads with Barbecue

Now, this is a match made in heaven. The smoky, tangy flavors of barbecue, whether ribs or pulled pork, beg for something with a bit of sweetness to balance them out. A honey-forward mead, like a wildflower mead, paired with barbecue hits all the right notes. The mead's sweetness complements the barbecue sauce, while its acidity helps cut through the richness of the meat. It was a revelation the first time I tried this pairing, and I haven't looked back since.

Spiced Meads with Lamb

Lamb has a slightly gamey flavor, and pairing it with a spiced mead made for an exciting combination. I once paired a spiced mead—heavy on cloves and allspice—with grilled lamb chops, and the complexity of the flavors was unforgettable. The mead's warmth wrapped around the lamb's richness, creating a dish that was both bold and comforting. The sweetness from the mead also balanced the gaminess of the lamb beautifully.

Pairing Mead with Dessert

Dessert pairings were where I first discovered how well mead complements food. The honey undertones naturally lean toward the sweet side, making mead a perfect match for desserts.

Sweet Meads with Cheesecake

The first time I tried a sweet mead with cheesecake, it was like a match made in heaven. The creamy, dense texture of the cheesecake was lifted by the honey-sweetness of the mead, creating a balance that was rich but not overwhelming. A mead made with orange blossom honey brought citrusy notes that played beautifully against the richness of the cheesecake, making each bite feel lighter.

Dry Meads with Apple Pie

Pairing a dry mead with the spiced sweetness of apple pie seemed counterintuitive at first, but it turned out to be one of my favorite combinations. The dry mead's crispness helped to cut through the buttery crust and caramelized apples, creating a harmony of sweet and tart flavors. It was as if the mead provided a palate-cleansing effect after each bite, making the next one even more enjoyable.

Spiced Meads with Dark Chocolate

This is a combination I always come back to when I want something decadent. The bitterness of dark chocolate and the warm, spiced notes of mead (think cinnamon, cardamom, and cloves) is a combination that feels both luxurious and comforting. The spices in the mead enhance the complex flavors of the chocolate, while its sweetness provides a gentle contrast to the chocolate's intensity.

CHAPTER NINETEEN

Pairing mead with cheese, meat, and dessert has been a rewarding journey of discovery for me. Mead's natural honey sweetness, along with its range of styles—dry, sweet, spiced, and fruit-infused—makes it an incredibly versatile beverage to pair with food. Whether it's a light goat cheese, a smoky barbecue, or a rich cheesecake, mead can elevate the flavors of a dish in unexpected and delightful ways.

Mead-Based Cocktails

While traditionally consumed on its own, mead's versatility allows it to serve as a unique and flavorful base for cocktails. Its wide range of flavor profiles—from dry to sweet, spiced to fruity—makes mead an excellent ingredient for mixologists looking to craft innovative drinks. Mead-based cocktails blend the ancient with the modern, combining traditional honey-wine flavors with contemporary mixology techniques to create refreshing and exciting drinks.

Why Use Mead in Cocktails?

Mead is a natural fit for cocktails because of its wide range of flavor profiles and its honey base. Mead can be light and refreshing, or rich and complex, depending on the ingredients used and the fermentation process. This versatility makes it an ideal ingredient for cocktails that require subtle sweetness, unique floral notes, or bold, spiced flavors.

- **Natural Sweetness**: Mead's honey base provides a natural sweetness that blends well with a variety of spirits and mixers. Unlike sugar or syrups, the sweetness of mead is more complex, often carrying floral, fruity, or earthy undertones depending on the type of honey used. This allows it to add depth and complexity to cocktails without

overwhelming them with sugar.
- **Flavor Flexibility**: Mead comes in a wide range of flavors, from dry and crisp to sweet and rich. There are also many meads infused with spices, herbs, and fruits, offering nearly endless possibilities for cocktail experimentation. This flexibility allows for creative combinations, whether you're looking to craft a light summer drink or a warming winter cocktail.
- **A Historical Twist**: Mead's ancient origins bring an element of tradition to modern mixology. Using mead in cocktails adds a layer of history and intrigue to any drink, creating a fusion of old-world craftsmanship and new-world creativity.

Popular Mead Cocktails and Their Variations

Mead can serve as the star ingredient or play a supporting role in cocktails. Here are some popular mead-based cocktails and variations that highlight the beverage's versatility.

1. Mead Mule

A refreshing twist on the classic Moscow Mule, the Mead Mule swaps vodka for mead, creating a drink that's lighter, more floral, and naturally sweet. This cocktail combines mead with ginger beer and lime juice for a zingy, refreshing drink perfect for warm weather or any time you crave something crisp and effervescent.

Ingredients:

- 2 oz mead (dry or semi-sweet)
- 4 oz ginger beer
- 1 oz lime juice
- Lime wedge and mint sprig for garnish

Instructions

Fill a mule mug with ice. Add the mead and lime juice, then top with ginger beer. Stir gently to combine. Garnish with a lime wedge and a sprig of mint. The honey sweetness of the mead complements the sharp bite of the ginger beer, while the lime adds a refreshing tang.

2. Mead Old Fashioned

For a sophisticated and spirit-forward drink, the Mead Old Fashioned offers a unique take on the traditional cocktail. Spiced or semi-sweet mead adds a layer of complexity to the classic whiskey-based cocktail, giving it a warming, honeyed finish.

Ingredients:

- 2 oz bourbon or rye whiskey
- 1 oz spiced mead
- 1 sugar cube
- 2 dashes of Angostura bitters
- Orange peel and cherry for garnish

Instructions

Muddle the sugar cube with the bitters in a mixing glass. Add the whiskey and mead, then stir with ice until well chilled. Strain into an old-fashioned glass over a large ice cube. Garnish with an orange peel and cherry. The mead adds a subtle sweetness and spice to the whiskey, rounding out the drink's bold flavors.

3. Mead Mojito

A fun, summery cocktail, the Mead Mojito swaps out rum for mead, resulting in a light, refreshing drink with a floral honey twist. The mint, lime, and mead combine to create a drink that's sweet, herbaceous, and tangy.

Ingredients

- 2 oz semi-sweet mead
- 1 oz lime juice
- 1/2 oz simple syrup (optional, depending on the sweetness of the mead)
- 6-8 fresh mint leaves
- Club soda
- Lime wedge and mint sprig for garnish

Instructions

Muddle the mint leaves and lime juice in a glass. Fill the glass with ice, add the mead, and top with club soda. Stir gently and garnish with a lime wedge and mint sprig. The honey from the mead pairs beautifully with the fresh mint and lime, making this a light and refreshing cocktail for hot summer days.

4. Mead Sangria

For a fruity and festive drink, mead sangria is a fantastic option. This cocktail combines mead with fresh fruit and a splash of brandy, creating a drink that's perfect for parties or outdoor gatherings. The fruit infuses the mead with its flavors, making this a delicious and customizable cocktail.

Ingredients

- 1 bottle of semi-sweet or fruit-infused mead
- 1/4 cup brandy
- 1/4 cup orange juice

- 1 orange, sliced
- 1 lemon, sliced
- 1 apple, chopped
- 1 cup of mixed berries (optional)

Instructions

Combine the mead, brandy, and orange juice in a large pitcher. Add the sliced fruits and berries, then refrigerate for at least two hours to allow the flavors to meld. Serve over ice. The fruity flavors of the mead pair perfectly with the fresh fruit, creating a sweet, refreshing cocktail that's easy to make and even easier to enjoy.

5. Mead Martini

For those who prefer a more spirit-forward drink, a mead martini offers a sweet, honeyed variation on the classic. This cocktail combines gin or vodka with mead, resulting in a smooth and elegant drink that's perfect for sipping.

Ingredients

- 2 oz gin or vodka
- 1 oz dry or semi-sweet mead
- Lemon twist for garnish

Instructions

Fill a mixing glass with ice and add the gin or vodka and mead. Stir until chilled, then strain into a martini glass. Garnish with a lemon twist. The mead adds a subtle sweetness and complexity to the martini, making it a refined yet approachable cocktail.

6. Spiced Mead Hot Toddy

Perfect for cold winter nights, a Spiced Mead Hot Toddy combines the comforting warmth of mead with the soothing flavors of cinnamon, cloves, and lemon. This cocktail is ideal for sipping by the fire or as a remedy for a cold.

Ingredients

- 1 oz whiskey or brandy
- 1 oz spiced mead
- 1 tablespoon honey
- 1/2 lemon, juiced
- 1 cinnamon stick
- Hot water

Instructions

In a mug, combine the whiskey or brandy, spiced mead, honey, and lemon juice. Top with hot water and stir to dissolve the honey. Garnish with a cinnamon stick. The combination of whiskey, mead, and warm spices creates a comforting, soothing drink perfect for chilly evenings.

Crafting Mead-Based Aperitifs and Digestifs

When I first began exploring mead, I was drawn to its ancient origins and unique honeyed flavor. But as I went deeper into the world of mead, I realized that this versatile drink has so much more to offer, especially when used as a base for aperitifs and digestifs. These traditional pre- and post-dinner drinks, meant to either stimulate the appetite or aid digestion, find a perfect partner in mead. Its broad flavor spectrum, from dry and herbaceous to sweet and spiced, makes it an excellent foundation for crafting sophisticated drinks. Let me walk you through my journey of creating mead-based aperitifs and digestifs, and how you can easily craft these at home.

Mead-Based Aperitifs

Aperitifs are typically light, refreshing, and designed to awaken the palate before a meal. Mead's subtle sweetness and complex flavors make it an excellent candidate for a variety of aperitif cocktails.

1. Dry Mead Spritz

One of the first aperitifs I tried crafting was a simple Dry Mead Spritz. I wanted something bubbly and refreshing, with just a hint of sweetness. The light, crisp nature of dry mead lends itself well to a spritz-style cocktail, where its delicate flavors can shine without being overpowering.

Ingredients:

- 3 oz dry mead
- 1 oz Aperol (or your favorite aperitif liqueur)
- 2 oz sparkling water
- Orange slice for garnish

Instructions

Fill a wine glass with ice and pour in the dry mead and Aperol. Top with sparkling water and stir gently to combine. Garnish with an orange slice. The dry mead gives this drink a crisp, honeyed undertone that pairs beautifully with the slightly bitter and citrusy notes of Aperol. It's a refreshing way to kick off a meal without overwhelming the palate.

2. Herb-Infused Mead Tonic

Another great aperitif I discovered was an Herb-Infused Mead Tonic. Inspired by traditional herbaceous cocktails, I wanted to create something that highlighted the subtle floral and herbal notes of mead. By infusing a semi-dry mead with herbs like thyme or rosemary, you can craft a drink that's light, refreshing, and aromatic.

Ingredients:

- 2 oz semi-dry mead
- 1 oz gin (optional for extra complexity)
- 1 oz tonic water
- Fresh thyme or rosemary sprig for garnish
- Lemon twist for garnish

Instructions

In a mixing glass, gently muddle the fresh thyme or rosemary to release the herbal aromas. Add ice, mead, gin (if using), and tonic water, then stir gently. Strain into a chilled glass and garnish with a lemon twist and an additional herb sprig. The semi-dry mead complements the tonic water's bitterness and the herbs' fresh, earthy notes, making for a delicate and refreshing drink that's perfect to enjoy before dinner.

3. Mead Negroni

One of my more adventurous experiments was a Mead Negroni. Traditionally, Negronis are quite strong, with bold flavors from gin, Campari, and sweet vermouth. But by substituting sweet vermouth with a spiced or fruity mead, you can create a mellower, more honeyed version of this classic aperitif.

Ingredients:

- 1 oz spiced or fruity mead
- 1 oz gin
- 1 oz Campari
- Orange peel for garnish

Instructions:

Combine the mead, gin, and Campari in a mixing glass filled with ice. Stir until well chilled, then strain into a glass over a large ice cube. Garnish with an orange peel. The sweetness of the mead balances the bitterness of the Campari, while the gin adds a sharp, botanical edge. The result is a smooth, slightly sweet take on a classic aperitif that's perfect for slowly sipping before dinner.

Mead-Based Digestifs

Digestifs are traditionally consumed after a meal, intended to help with digestion and provide a satisfying end to the evening. Mead's rich honey notes, especially when paired with warming spices or bold flavors, make it a wonderful base for digestif-style drinks.

1. Spiced Mead Toddy

One of my go-to mead digestifs is a Spiced Mead Toddy. There's something incredibly comforting about a warm drink after dinner, and the rich spices in a mead toddy feel like a hug in a glass. It's the perfect digestif for a chilly evening, when you want something soothing and a bit indulgent.

Ingredients:

- 2 oz spiced mead
- 1 oz whiskey or brandy
- 1 tablespoon honey
- 1 cinnamon stick
- Lemon slice for garnish
- Hot water

Instructions:

In a mug, combine the mead, whiskey, and honey. Add a cinnamon stick and fill the mug with hot water, stirring to dissolve the honey. Garnish with a lemon slice. The combination of spiced mead and whiskey creates a drink that's both warming and soothing. The honey sweetness lingers on the palate, while the cinnamon adds an aromatic touch, making it the perfect way to relax after a hearty meal.

2. Honeyed Mead Old Fashioned

For those who enjoy a stronger, spirit-forward digestif, a Honeyed Mead Old Fashioned is an excellent option. The addition of sweet mead to this classic cocktail adds a layer of complexity and richness, making it an indulgent treat after dinner.

Ingredients:

- 1 oz bourbon or rye whiskey
- 1 oz sweet mead
- 2 dashes of Angostura bitters
- 1 sugar cube
- Orange peel for garnish

Instructions

Muddle the sugar cube and bitters in a glass. Add the whiskey and mead, and stir with ice until well chilled. Garnish with an orange peel. The mead's sweetness tempers the bourbon's boldness, while the bitters add depth. This digestif is smooth and warming, perfect for sipping slowly as you wind down the evening.

3. Fruit Mead Cordial

Another delightful digestif is a Fruit Mead Cordial. If you have a bottle of fruit-infused mead on hand, like blackberry or raspberry mead, it can be turned into a simple yet elegant cordial. The fruity flavors paired with the honey base create a luscious, velvety drink that's perfect for a post-dinner treat.

Ingredients:

- 2 oz fruit-infused mead (blackberry or raspberry work well)
- 1 oz brandy
- Splash of soda water
- Fresh berries for garnish

Instructions:

In a glass filled with ice, combine the fruit mead and brandy. Stir gently, then top with a splash of soda water for a bit of effervescence. Garnish with fresh berries. The rich, fruity notes of the mead complement the smoothness of the brandy, creating a digestif that's both refreshing and indulgent. The soda water lightens the drink, making it a lovely option for after dinner.

20

CONCLUSION.

As we conclude *THE MEAD MAKING BIBLE*, I extend my sincere gratitude to you, the reader, for joining me on this exploration of mead-making. It has been a privilege to guide you through the intricate processes, rich history, and modern innovations that define this remarkable craft. Your engagement and enthusiasm have been invaluable, and I deeply appreciate your dedication to learning and mastering this ancient art.

This book has aimed to provide a thorough foundation, from the fundamental principles to advanced techniques, with the hope of inspiring both novice and seasoned mead-makers alike. The knowledge and practices shared herein are designed not just to instruct but to empower you to experiment, innovate, and personalize your mead-making journey.

As you move forward, I encourage you to continue exploring and refining your skills. The world of mead is vast and ever-evolving, and your own contributions will undoubtedly add to its rich tapestry. The principles and recipes presented in this book are a starting point, and your individual creativity will be the key to unlocking new and exciting possibilities.

For those embarking on the journey of mead-making, here are a few key pieces of advice to set you up for success:

CONCLUSION.

1. **Start Simple:** Begin with a basic recipe to familiarize yourself with the process before experimenting with complex flavors or techniques. A simple honey, water, and yeast mead will help you understand the fundamentals.
2. **Quality Ingredients:** Use high-quality honey, as it's the primary source of flavor in mead. Fresh, unprocessed honey will yield the best results.
3. **Sanitation is Key:** Ensure that all equipment is thoroughly cleaned and sanitized. Contaminants can spoil your mead and derail your efforts.
4. **Patience is Essential:** Mead-making takes time. Allow your mead to ferment, age, and develop its flavors fully. Rushing the process can lead to disappointing results.
5. **Record Everything:** Keep detailed notes on your recipes, procedures, and observations. This will help you track what works and what doesn't, leading to better outcomes in future batches.
6. **Taste and Adjust:** Regularly sample your mead throughout the fermentation process to understand its development. This can help you make informed adjustments to your recipe.
7. **Learn Continuously:** Join mead-making communities, attend workshops, and read extensively. The more you learn, the better your mead will become.

Approach mead-making with curiosity and a willingness to learn. Each batch is an opportunity to refine your skills and deepen your appreciation for this ancient craft.

Thank you once again for your support and interest. I look forward to continuing this journey with you, sharing further insights, developments, and discoveries in the world of mead-making. Your commitment to this craft is a testament to the enduring allure and significance of mead, and I am honored to be part of your mead-making experience.

www.ingramcontent.com/pod-product-compliance
Lightning Source LLC
Chambersburg PA
CBHW052144220526
45471CB00004B/1513